Supersymmetry

Squarks, Photinos, and the Unveiling of the Ultimate Laws of Nature

·   ·

GORDON KANE

§
HELIX BOOKS

PERSEUS PUBLISHING
Cambridge, Massachusetts

Many of the designations used by manufacturers and sellers to distinguish their products are claimed as trademarks. Where those designations appear in this book and Perseus Publishing was aware of a trademark claim, the designations have been printed in initial capital letters.

A CIP record for this book is available from the Library of Congress.
ISBN: 0–7382–0489–7

Perseus Publishing is a member of the Perseus Books Group.

Find us on the World Wide Web at http://www.perseuspublishing.com

Text design by Jeff Williams
Set in 11-point Minion by Perseus Publishing Services

1 2 3 4 5 6 7 8 9 10—03 02 01
First paperback printing, June 2001

Perseus Publishing books are available at special discounts for bulk purchases in the U.S. by corporations, institutions, and other organizations. For more information, please contact the Special Markets Department at the Perseus Books Group, 11 Cambridge Center, Cambridge, MA 02142, or call (617) 252-5298.

To Hal, Mollie, David, and Noah

Contents

The physics of the Planck scale • Effective theories replace renormalization • The human scales

Foreword

Looking back at century's end, it is stunning to think how our understanding of physics has changed in the last hundred years. The great insights of the early part of the century were of course Relativity Theory and Quantum Mechanics. We learned in Einstein's Special Relativity theory of the strange behavior of fast moving objects, and in his even more surprising General Relativity we learned to reinterpret gravity in terms of the curvature of space and time caused by matter. As for Quantum Mechanics, it taught us that fact is far more wondrous than fiction in the atomic world.

Special Relativity and Quantum Mechanics were fused in Quantum Field Theory, whose most remarkable prediction—verified experimentally in cosmic rays around 1930—is the existence of "antimatter." Quantum Field Theory is a very difficult theory to understand even for specialists; trying to understand it has occupied the attention of many leading physicists for generations.

The last fifty years have been an amazing period of experimental discoveries and surprises, including "strange particles," the breaking of symmetry between left and right and between past and future, neutrinos, quarks, and more. Drawing on this material, theoretical physicists have been able, in the Quantum Field Theory framework, to construct the Standard Model of particle physics, which puts under one roof most of what we know about fundamental physics. It describes in one framework electricity and magnetism, the weak force responsible among other things for nuclear beta decay, and the nuclear force.

Is this journey of discovery nearing an end? Or will the next half century be a period of surprises and discoveries rivaling those of the past? The questions we can ask today are as exciting as any in the past, and at least some of the answers can be found in the coming period if we stay the course.

Just in the last few months, newspapers have been filled with exciting accounts of recent and forthcoming experiments testing the strange properties of neutrinos, and perhaps showing that the "little neutral particle" of Fermi does have a tiny but nonzero mass after all. Astronomers have unraveled new and challenging hints that General Relativity may need to be corrected by adding Einstein's "cosmological constant"—the energy of the vacuum. Novel and inventive dark matter searches are probing the invisible stuff of the universe. Satellite probes of fluctuations in the leftover radiation from the big bang are likely, in the next few years, to challenge our understanding of the large scale structure of the universe.

But one of the biggest adventures of all is the search for "supersymmetry." Supersymmetry is the framework in which theoretical physicists have sought to answer some of the questions left open by the Standard Model of particle physics. The Standard Model, for example, does not explain the particle masses. If particles had the huge masses allowed by the Standard Model, the universe would be a completely different place. There would be no stars, planets, or people, since any collection of more than a handful of elementary particles would collapse into a Black Hole. Subtle mysteries of modern physics—like spacetime curvature, Black Holes, and quantum gravity—would be obvious in everyday life, except that there would be no everyday life.

Supersymmetry, if it holds in nature, is part of the quantum structure of space and time. In everyday life, we measure space and time by numbers, "It is now three o'clock, the elevation is two hundred meters above sea level," and so on. Numbers are classical concepts, known to humans since long before Quantum Mechanics was developed in the early twentieth century. The discovery of Quantum Mechanics changed our understanding of almost everything in physics, but our basic way of thinking about space and time has not yet been affected.

Showing that nature is supersymmetric would change that, by revealing a quantum dimension of space and time, not measurable by ordinary numbers. This quantum dimension would be manifested in the existence of new elementary particles, which would be produced in accelerators and whose behavior would be governed by supersymmetric laws. Experimental clues suggest that the energy required to produce the new particles is not much higher than that of present accelerators. If supersymmetry plays the role in physics that we suspect it does, then it is very likely to be discovered by the next generation of particle accelerators, either at Fermilab in Batavia, Illinois, or at CERN in Geneva, Switzerland.

When Einstein introduced Special Relativity in 1905 and then General Relativity in 1915, Quantum Mechanics was still largely in the future, and Einstein assumed that space and time can be measured by ordinary numbers. Einstein's conception of space and time has been adequate for discoveries made until the present, but discovery of supersymmetry would begin a reworking of Einstein's ideas in the light of Quantum Mechanics.

Discovery of supersymmetry would be one of the real milestones in physics, made even more exciting by its close links to still more ambitious theoretical ideas. Indeed, supersymmetry is one of the basic requirements of "string theory," which is the framework in which theoretical physicists have had some success in unifying gravity with the rest of the elementary particle forces. Discovery of supersymmetry would surely give string theory an enormous boost.

The search for supersymmetry is one of the great dramas in present-day physics. Hopefully, the present book will introduce a wider audience to this ongoing drama!

Edward Witten
Princeton, New Jersey
June 30, 1999

Preface

If you take a little trouble, you will attain to a thorough understanding of these truths. For one thing will be illuminated by another, and eyeless night will not rob you of your road till you have looked into the heart of nature's darkest mysteries. So surely will facts throw light upon facts.
    —Lucretius, *On the Nature of the Universe*
    (Translated by R. E. Latham, Penguin Books)

Most people realize that anyone who is interested in how an old-fashioned watch works can get a good idea of what is happening inside. Few people, however, realize that physicists now have a similarly clear image of the mechanisms of the subatomic universe—the stuff that makes the world run. That image is formulated in what we call the Standard Model of particle physics. It is a *description* of the underlying structure of the universe. Someone who probes and studies a watch not only can *describe* the workings of that watch but also can say *why* the watch works—why this cog moving that one at a given ratio mimics the progress of time. Physicists, too, are increasingly able to peer into the workings of the universe and say *why* the ingredients they study are able to create and sustain what we know as nature.

A nice way to learn more about a watch is to see a watchmaker disassemble and reassemble one, and a nice way to learn more about nature is to go for a walk with a naturalist. This book is meant as a leisurely walk that can be enjoyed by anyone with the curiosity to come along and observe the particles and their behavior. We will stroll not only in the known territory of the Standard Model but also along the frontier of topics where breakthroughs into even more remote regions may soon occur. For various practical and theoretical reasons, many particle physicists think that the next major discovery will be direct evidence for the property called *supersymmetry*.

Those reasons, and the implications if direct evidence of supersymmetry is indeed observed, are much of what this book is about.

The first phase of the age-old search for understanding how the physical world works has been brought to a successful close in recent years with the development and testing of the Standard Model of particle physics. The Standard Model (summarized in Chapter 2) gives a comprehensive description of the basic particles and forces of nature and of how all of the physical phenomena we see can be described. It contains the underlying principles of all the behavior of protons, nuclei, atoms, molecules, condensed matter, stars, and more. The Standard Model has explained much that was not understood before; it has made hundreds of successful predictions, including many dramatic ones; and there are no phenomena in its domain (Chapter 3) that are not explained (though some calculations are too complicated to carry through). The Standard Model has loose ends (such as the "Higgs physics," Chapter 7), but they don't affect most of its explanations and tests.

If the Standard Model describes the world successfully, how can there be physics beyond it, such as supersymmetry? There are two reasons. First, the Standard Model does not explain aspects of the study of the large-scale universe, cosmology. For example, the Standard Model cannot explain why the universe is made of matter and not antimatter (Chapter 8), nor can it explain what the dark matter of the universe is (Chapter 6). Supersymmetry suggests explanations for both of these mysteries. Second, the boundaries of physics have been changing. Now scientists ask not only how the world works (a question the Standard Model answers) but why it works that way (a question the Standard Model cannot answer). Einstein asked "why" earlier in the century, but only in the past decade or so have the "why" questions become normal scientific research in particle physics, rather than philosophical afterthoughts. One ambitious approach to "why" is known as *string theory* (Chapter 9), which is formulated in an eleven-dimensional world. Work on string theory has proceeded so far by study of the theory itself, rather than via the historically fruitful interplay of experiment and theory. This approach has led to significant and exciting progress; if it succeeds we will all be delighted. As Edward Witten remarks in his Foreword to this book, string theory predicts that nature should be supersymmetric.

Supersymmetry is a surprising and subtle idea—the idea that the equations representing the basic laws of nature don't change if certain particles in the equations are interchanged with one another. Just as a square on a piece of paper looks the same if you rotate it by 90 degrees, the equations that physicists have found to describe nature often do not change when certain operations are performed on them. When that happens, the equations are said to have a symmetry. Supersymmetry is such a proposed symmetry—the "super" is included in its name because this symmetry (Chapter 4) is more surprising and more hidden from everyday view than previously discovered symmetries. It turns out that the idea has remarkable consequences for explaining aspects of the world that the Standard Model cannot explain, particularly the Higgs physics; they are described in Chapters 4–8. The most important implication may be that supersymmetry can provide a window that enables us to look at the minute world of string theory from our full-size world, so that experiment can provide guidance to help formulate string theory, and so that the predictions of string theory can be tested. Supersymmetry ushers in the second phase of the search for understanding.

Supersymmetry is still an idea as this book is being written (mid-1999). There is considerable indirect evidence that it is a property of the laws of nature, but the confirming direct evidence is not yet in place. That is not an argument against nature being supersymmetric; rather, the accelerator facilities that could confirm it are just beginning to cover the region where the signals could appear (Chapter 5). I have tried to present the material in this book in such a way that it will remain valid and interesting after the superpartners and Higgs bosons predicted by supersymmetry are found. When we have positive signals, the focus can be sharpened, but the explanations that supersymmetry can provide, the way it can connect with string theory, and how we recognize and test it are likely to be very close to what is presented here.

If the world we live in does exhibit the property called supersymmetry, even though it has been hidden from our view until now, we will have a systematic way to peer at the most basic law(s) that govern nature and our universe. Without supersymmetry that may not be possible. Though there is considerable indirect evidence that the world is indeed supersymmetric, this is not yet certain. It is worth a lot of effort to find out.

A number of people have enriched this book. I am very grateful to my most relentless editor, my wife Lois, who contributed greatly to the book's intelligibility; to Jim Wells and Lisa Everett for many very helpful suggestions; and to Kate Logan for extensive assistance, particularly with the figures. I appreciate very much the encouragement I have received from Perseus Books, comments from Steve Mrenna, help from Judy Jackson in obtaining Fermilab photos, and help from Jane Nachtman, Andrei Nomerotski, Daniel Treille, Jianming Qian, and Saul Youssef in obtaining pictures of events.

Supersymmetry

Where Do We Come From? What Are We? Where Are We Going?

Paul Gauguin titled what he thought would be his last painting "Where do we come from? What are we? Where are we going?" In his writings he mentioned its "enormous mathematical faults" and how it was "all done from imagination." Gauguin's reflections remind me of where scientists are today in the search for a complete understanding of our universe. Scientists work with mathematical constructions and imagine hypotheses while trying to grasp where we come from, what we are, and where we are going—or, more concretely, while trying to establish why there is a universe, how and why it works the way it does, what we are made of, and how inanimate matter can give rise to conscious, thinking people.

Every culture has asked these questions in some form and has followed some approach to provide answers. The approach that we call *science* has led to a remarkable set of results and answers to some of these questions, because it developed a method to study the natural world. The scientific method began with the Ionian Greeks over 2500 years ago and began to provide reliable knowledge about the world with the work of Galileo and Kepler about 400 years ago. Science makes progress by combining imagination with experimental results—by insisting on evidence.

FIGURE 1.1.

More than one physicist, attracted by the title, has a reproduction of Gauguin's painting. But when I look at it (Figure 1.1), I don't see the answers that Gauguin perhaps had in mind, because the painting is his personal approach to those questions we all ponder. Science, on the other hand, allows many to search for answers together and to interpret the answers for whoever is interested. I hope this book will help do that for the reader. Science poses the same questions Gauguin and other artists ask. Its aim is to understand what lies behind the verb form *to be*. Though some believe otherwise, this science is not the opposite of the humanities, though it may be less readily portrayed in verbal and visual images. Quarks can't really be represented by curly beards or white togas, electromagnetic fields can't be shown as pudgy babies with wings and bows and arrows. Equations and their solutions are the representational images of the universe's structure; the circumference of a circle and the parabola described by the path of a cannon ball are both precise and beautiful images of aspects of nature. If someday we have a complete set of equations, perhaps unified into one primary equation, we will have a complete mathematical image of the universe. Then we will be able to convert that to a verbal image.

Today we are at a stage where there is one main idea about the next experimentally accessible step toward understanding the basic laws that govern the universe, but it is very hard, for practical reasons, to get the evidence we need in order to learn whether the idea is correct. This book focuses on that idea, which is called supersymmetry. There is already indirect evidence (it will be described in later chapters) that supersymmetry is part of a correct description of nature. If we understand supersymmetry and its implications correctly, direct experimental evidence for supersymmetry will be found in the next few years—possibly soon after this book is published. As we will see, supersymmetry is important not only as a possible previously unknown part of nature, but in addition because it should allow us to probe the ultimate laws of nature much more directly.

This first chapter is meant to explain what this book is about, and describe its assumptions and goals. It can be difficult to understand how science works, how it progresses, and how scientists working in an area become convinced an accurate description of nature (or the universe, or the world—in this book we'll use these words essentially interchangeably) has been formulated. It can also be difficult to understand the results. The next pages are an effort to prevent misunderstandings and lead us smoothly into our subject.

TO UNDERSTAND NATURE WE NEED TO KNOW ABOUT PARTICLES, FORCES, AND RULES

In order to understand the natural world, we have to know at least three things. As we probe the world we find that it consists of particles, so we have to know what the basic particles are. Over two millennia ago, some Greeks correctly reasoned that the wonderful complexity of the world we see could be explained if everything were composed of a number of basic, irreducible constituents (particles). It wasn't until the past century, however, that we developed the techniques needed to test ideas about these particles and to establish their existence and properties; Chapter 2 will describe the results.

The particles interact to form all the structures of our world, so we have to know how they interact as well: what forces or interactions affect them. But even if we know all about the particles and forces, we cannot explain anything unless we also know nature's rules, and have mathematical representations of them, so that we can work out the behavior of the particles under the influence of the forces. For example, even if we know that two particles will attract each other and fall toward each other because of the gravitational force, we don't know how energetic a collision they will undergo unless we have an equation (a rule) by which to calculate their speed. Nature's rules apply for all particles and interactions. The first rules were written down by Issac Newton. Today his rules and others are integrated into two comprehensive rules: Einstein's so-called special relativity and the quantum theory. These rules will not play much of a detailed or visible role in this book, and you don't need to know how they work to understand the book. The important thing is that they are there in the background, as well-established and well-tested methods to calculate how particles behave when they interact through a given set of forces.

One can ask whether our formulation of quantum theory and special relativity is likely to be extended or modified as progress is made toward an ultimate theory of nature. That is unlikely, at least for all practical purposes. The equations and algorithms that represent nature's rules have been extremely well tested in a variety of situations, but the reasons to believe they will remain valid are even stronger than the explicit tests. The equations and algorithms are part of a mathematical theory that forms a coherent structure. If any part of that theory were changed, the change would propagate through to other parts, and would be likely to lead to untenable changes in some well-tested part. (We will see later in the book that there might be some room to change the formulations for extremely high-energy interactions, though there is no reason to think such changes will occur.) To fully understand the world, then, we will have to understand not only what nature's rules are but also why they are the rules. The effort to understand this is barely beginning to be a subject of research. But we do know enough to be confident that for purposes of formulating and understanding and testing supersymmetry, our present knowledge of the rules is satisfactory.

As I noted above, for this book the reader does not need to know much about nature's two basic rules beyond the fact that they are there, but it's worthwhile to describe them a little. The constraints of special relativity follow from two simple postulates, which can basically be stated as follows: (1) The laws of nature are the same regardless of where they are formulated and tested. (2) The speed of light in vacuum (denoted by c) is the same regardless of the conditions under which it is measured. The first principle is obvious; it says what it seems to say—that if you work out the laws of nature on earth, on another planet across the galaxy, on a spaceship, or anywhere, you will get the same results. The second is not so obvious, but the fact that the speed of light in vacuum is always the same has been extremely well tested by many approaches, both directly and by examination of the implications of special relativity.

The word *relativity* in this theory's name is misleading and unfortunate, because the essence of special relativity is that two things are absolute, not relative at all. The name stems from an implication of the theory: that the outcome of some experiment can be different if the experiment is carried out in laboratories that are in relative motion (such as one on earth's surface and the other in an airplane moving at constant speed overhead). However, the theory goes on to show that when the effects of the motion are included, even for experiments in relative motion the resulting descriptions of the results become the same.

We are interested here in special relativity mainly because it limits the form that a valid theory can take. It's a powerful constraint—for example, Newton's laws had to be reformulated because they did not originally obey the constraints of special relativity. Whenever we need to say that a theory does (or does not) satisfy the constraints of special relativity, we will use the physics jargon and say the theory is "relativistically invariant" or that it satisfies "relativistic invariance" constraints. Special relativity was fully formulated by Einstein in 1905. Its validity was tested both theoretically (it had to be consistent with all verified descriptions of nature) and experimentally over several decades. It is still being tested whenever new technologies become available.

The other part of the basic rules, quantum theory, was formulated between 1913 and 1927 by several people. For this book we do not need to know much about how quantum theory works, only that it is there and that it tells us how to calculate the behavior of particles if we know the forces that affect them. Later in the book we'll learn a few properties of quantum theory that we need for specific purposes. Special relativity and quantum theory have been successfully combined into a "relativistic quantum theory." Whenever that phrase appears in this book, it means that the ideas under discussion have been successfully formulated to obey simultaneously the rules represented by special relativity and quantum theory.

Before about 1965 we knew very little about what basic particles were the constituents of matter. We knew what forces existed, but not how they worked to shape the world. By the end of the 1980s, we had learned what the basic constituents of matter are, and we understood how particles and forces function to make our world. That body of knowledge is called the Standard Model of particle physics; it is the subject of the next chapter. It provides a well-tested description of how our world works. I'll normally refer to this as the Standard Model, but the qualifier "of particle physics" should always be assumed. If nature is supersymmetric, then the Standard Model will be *extended* to become the Supersymmetric Standard Model. The Standard Model will not be wrong but will simply become a part of a more complete description of nature. That is the way science progresses. Once an area is well tested and established, it is not dropped as the description of nature broadens but is extended and integrated into the new picture.

The Standard Model is not a model in any conventional sense of that word; rather, it is the most complete mathematical theory ever developed. For physicists, *theory* does not mean what it might in everyday usage. Most generally, a theory is a principle or set of principles that imply properties of the natural world. In physics a theory typically takes the form of a well-defined set of equations that expresses relationships among some symbols. The symbols represent parts of the natural world. The equations can be solved to learn the behavior of the quantities, and thus the predicted behavior of the parts of the world the quantities represent. For example, consider Einstein's famous equation $E = mc^2$, a consequence of special relativity. Here the symbols are E, m, and c. E represents an amount of energy, m a mass, and c the speed of light. If an amount of mass m is converted into energy, this equation tells us how much energy is obtained. Solving this equation for E is easy—we just multiply m by c^2. In general, solving equations can be much harder. (The universal validity of the relationship $E = mc^2$ is one of the many reasons why we are confident that special relativity theory is correct. It is tested daily at colliders and in nuclear reactors.)

In science the use of the term *theory* carries a loose implication that the predictions are largely tested and verified. When scientists start to study an area of the world, they first make models to guide their thinking, suggest experiments that might be relevant to making progress, and allow quantitative predictions. Models usually begin as limited mathematical descriptions of how some aspects of the world behave. If they work well, more phenomena are added. Later, one improved model turns out to describe nature rather well, and it is often named the standard model. Even later, a version of that model is so robust and well tested that it becomes the theory of that area. But it is already called the "standard model," so the name stays, even though it is really not a model any more. We acknowledge that development in a limp way by capitalizing *Standard Model*. In everyday usage, *theory* means something very different: a kind of vague idea about how to explain something. We might say, for example, "One theory about the falling crime rate is...." The term *model* is used similarly in science and in everyday language, at least until a model is found that successfully describes nature, but the use of *theory* is very different and can cause confusion. In this book I will of course stick to the scientific use only. The correspondence between the theory and its equations and the structure of the physical system it represents is so exact that physicists conventionally (and perhaps confusingly) regard them as interchangeable in writing and discussions.

RESEARCH IN PROGRESS (RIP)

One of the main sources of confusion and misunderstanding of science by many (to say nothing of some journalists, philosophers, historians, sociologists, and even scientists themselves), is the failure to distinguish carefully between those areas of science that are conceptually and experimentally well established and those that are still developing. As science develops, it becomes possible to study some previously inaccessible part of the natural world. Most often a fresh area opens up because of technological innovations that become available—before the microscope was invented, it was not possible to study phenomena smaller than what we could see with the unaided eye. The opportunity to study new areas depends sometimes on the past successes of science itself, since subfields build on earlier ones, and sometimes on the introduction of new concepts. Eventually the foundations of a description of that aspect of the natural world are worked out and experimentally verified. This has occurred many times—for example, with optics, electromagnetism, thermodynamics, atomic physics, the Standard Model of particle physics, and other areas.

As a subfield is being studied and worked out, models and explanations and proposed ideas repeatedly change. Experiments often don't work correctly the first time they are done, but experiments are constantly redone and improved, so questions can be settled by improved experiments rather than argumentation. Experiments are always done with the variables that the equations depend on restricted to some specific range: variables such as velocity, astronomical distance, resolution of a microscope, and so on. Historically, once a theory is successfully tested in a given range of the variables it depends on, it will always work in that domain of the variables. However, if the theory is extended to new values of the variables—faster speeds, smaller or larger distances, and so on—it may work successfully in the new domain, or it may not. When it does not, eventually a new theory will be found that works in the larger domain. That new theory will "reduce to" the older one when the values of the variables are restricted to the older range. The older theory is not "wrong"; it is extended to become a part of the more general one. Eventually the entire range of variables may be covered. For example, the description of motion has now been tested for all possible speeds, from zero to the speed of light, and it will not change in the future. All the well-known examples of theories or rules once thought to be valid and later found to be inadequate fit this general picture.

Once upon a time there was no method called science. We have had to learn how to do science at the same time as we were learning the results of science. The process I have just described was not originally understood. For example, Newtonian science was extremely successful, and people naturally assumed it held for the whole universe. Only later did they realize that it was only approximately valid; it held when velocities and masses were not too large, and it failed for systems of atomic size. By the 1930s, physicists had finally realized that every theory had to be tested again whenever it was extrapolated beyond the range of variables where it was known to be valid.

Even though the foundations of many areas of physics (and other subfields of science) are in place, this does not mean they are no longer active areas of research. On the contrary, many people become excited about working out the implications of the new foundations. Having the foundations in place means that the basic equations that govern behavior in that subfield are known. But as we have already discussed briefly and as we will encounter again, finding the solutions of the equations can be difficult. Having the solutions for one system does not guarantee that we will have them for the next, so most subfields continue to pose interesting problems long after their basic principles are understood. When the theory is complicated, new predictions and results can emerge from further study or experimentation, even after a long time. To put it differently, most areas generate many applications once their basic principles are understood, and will continue to do so.

For our purposes, the main distinction to keep firmly in mind is that between subfields for which the foundations are in place and those where research is in progress. To alert the reader to areas where *Research is In Progress*, we will occasionally use one of the few acronyms in the book, RIP. For RIP, even though the current ideas have some experimental support and are esthetically very attractive, they may turn out to be partly or completely wrong. Of course, human error occurs frequently, leading to reported experiments or calculations that are wrong, but such errors are quickly caught and fixed if they matter; that is not what *wrong* means here. RIP refers to a deeper situation, where the very laws that describe nature's behavior are not yet discovered. RIP such as supersymmetry and super-string theory could simply turn out not to be how nature behaves. That is very different from the Standard Model, which is now known to describe how nature does behave in its domain. The Standard Model will be extended to be part of a larger theory, but in its domain it will not change. It could happen that we talk of some areas differently or even interpret their implications differently after they are extended, but in the Standard Model's domain, any new version will be technically equivalent to the established one.

Today's graduate students in physics are taught more about relativity than Einstein knew. Thousands of scientists use quantum theory daily. Once an area is mature, ways to explain it are developed. Writing or talking about RIP areas is challenging, mainly because the final answers are not yet in place, or at least are not confirmed and tested, but also because there has not been time or inclination to develop nontechnical analogies and explanations. If and when supersymmetry and the primary theory it moves us toward are in place, they will be easier to explain than they are now.

Part of the general problem in communicating science results to the general public is that the newsworthy results are the new ones, so they are usually RIP and thus are subject to all of the uncertainties of RIP. The results may be modified later as experiments improve and ideas continue to be tested. The media rarely stress the self-correcting and tentative nature of the research that exists until the subfield is understood.

For the nonscientist, the most important and interesting aspects of any area where research is in progress are not the details but what questions are under study and what kinds of answers are being considered. Subjects that previously were the domain of philosophy have become accessible to science as new techniques have been invented and more of the world has come to be understood. The answers currently favored in RIP areas may change (or they may not). Historically, once a subfield became a scientific research area, the questions were eventually answered. If history is a useful guide, that typically takes a few decades. As new data appear, and as ideas are confirmed or improved, those who follow such developments will be able to understand qualitatively how the field progresses. Some RIP areas are so speculative—so likely to survive only in dramatically modified forms—that articles and books about them may confuse general readers more than they inform them. Supersymmetry is now a sufficiently mature area, and sufficiently close to confirmation if it is indeed a part of the correct description of nature, that a wider understanding of its content and implications is both possible and worthwhile.

EQUATIONS?

Sometimes authors of science books meant for a general audience suggest that equations frighten readers away. That's probably true for the technical equations that scientists use in their work. But one of the remarkable things about equations is that you can start from an input you understand and agree is established, follow a few correct steps, and end up with a surprising result that you would never have believed if you had followed only a verbal argument. Simple equations, used as metaphors, can sometimes clarify ideas more than descriptions in words—that's mainly how we'll use the few equations in this book. Similarly, the pedigrees and characteristics of the numerous particles we meet should not cause concern. They are there if you like them, but all the basic ideas and the plot will be clear even if you don't remember the names of the characters (such as muons, neutralinos, etc.).

There is a major distinction between the properties of an equation and its solutions. We think of the theory as an equation (or several equations). On the other hand, our world is described by the *solutions* of the equation. That's how it always is in science. The principles are embodied in equations. The actual aspects of the world are described by solving the equations and finding out how the solutions behave. It can easily happen that the equations have some properties that a given solution does not have. This concept is unfamiliar to most readers, so let's explain it with an example that does illustrate this idea, although it does not correspond to any real situation. However, as we often do with examples and models in science, let's use one that has similarities to a real puzzle: why the world contains three particles that apparently are the same except for their masses. Even though the equations of the example aren't realistic, they illustrate how the true equations might behave.

Suppose a theory tells us that the equation relating the masses of three particles (call them the electron, the muon, and the tau—we'll see why in the next chapter) is

$$EMT = 64$$

Here E stands for the mass of the electron, M for the mass of the muon, and T for the mass of the tau. The equation just says multiply the three masses and you get 64. (We're ignoring units here.) This equation is totally symmetric—if you exchange any of the masses, you get the same equation back. Having the laws be symmetric is generally very desirable. If you didn't think about it, you might guess that the individual values of the masses would then all be equal because they come from solving a symmetric equation.

Let's list solutions of this equation. For simplicity, consider only solutions for which E, M, and T are integers. There are several sets of three numbers whose product is 64. One solution has $E = M = T = 4$, since $4 \times 4 \times 4 = 64$. For this solution all the masses are indeed equal so it is a symmetric solution. But another solution has $E = 1$, $M = 2$, and $T = 32$, which again multiply to 64. There are lots more—for example, (E, M, T) can be $(1, 1, 64)$; $(1, 4, 16)$; $(1, 8, 8)$; $(2, 2, 16)$; or $(2, 4, 8)$. In each case, just multiply and you get 64. (In our example E, M, and T represent particles that are the same except for their masses, so we define E to be the lightest and T the heaviest.) If all of them are solutions, how do we know which solution should actually describe nature? In this "toy" case, we can solve the equations to find all the solutions, but in the real world, solving the equations is often extremely difficult. And solving the equations is not enough, because we have to know how nature ended up being described by one of the solutions and not the others.

To show the possible power of data, suppose that the masses weren't well measured, but we knew from experiment that none of the three masses was equal to any other—then only three of the solutions could be correct. If we also knew from the data that none of the masses was more than about five times heavier than the others, we would be led to a unique solution, $(2, 4, 8)$.

This instructive little example illustrates several things. The principles and laws are embodied in equations that often are highly symmetric (for example, supersymmetric). The world (that is, the particles and how they interact) is described by the solutions to those equations. The solutions do not need to show the symmetry of the equations, and in general they do not. The symmetry of the basic law is hidden if we can observe only the solutions. Some of the strongest challenges scientists face in going from the world we observe to learning the laws that govern it arise because the laws have hidden symmetries that are not apparent in the world (such as the particle interchange symmetry of the Standard Model, described in Chapter 2). As we will see, supersymmetry itself is thought to be a well-hidden symmetry. Thus even if we have the basic laws, we are unlikely to be able to figure out which solution describes the world unless we have some relevant data from experiment and observation.

To be fair, though, it could happen that the theory is so powerful that it successfully picks out the correct solution, at least in principle. Let's extend our example. Suppose the theory produces a second equation, $E + M + T = 14$, in addition to $EMT = 64$. Both equations are entirely symmetric, but the only solution that simultaneously satisfies both equations is the one we found above, $E = 2$, $M = 4$, $T = 8$, and we did not need any data to learn that. Note that the two fully symmetric equations have one common unique solution, and it does not show the symmetry—perhaps our world is like that. We still need data to test whether any solution we find is the actual one that describes nature.

Prediction, Postdiction, and Testing

The goal of science is to achieve understanding. One method or tool that scientists use to move toward understanding is to make predictions and test them. Ideas and theories generate testable implications about the world. Tests of predictions can have three kinds of outcomes. If the prediction is verified, our confidence in the ideas that led to the prediction is strengthened. If the prediction is wrong, the theory must be modified or discarded. But with RIP, the situation is frequently more subtle. Often a prediction can be made in principle but depends on some quantities (such as masses) that are incompletely known. Then an iterative process occurs. Encouraging results entice more people into measuring or calculating poorly known quantities more accurately. The prediction and its tests are refined over time. Eventually the prediction is tested.

Another subtlety is that it can be meaningful to "predict" something that is already known, such as the mass of the electron, the existence of the force of gravity, or that we live in three space dimensions. Sometimes such predictions are called *postdictions*. They are meaningful because they occur in a theory that uniquely requires such an outcome. Such results can be powerful tests of the theory even though they were known before the theory was formulated. In other cases, a theory can address an issue such as the mass of the electron (whereas earlier theories could not even answer the question in principle) but be unable to provide a definite numerical answer or prediction because some input information is not yet known. In this case, it may be possible to predict that the outcome is in a certain range that includes the known result. This outcome is very encouraging compared to predicting the wrong range or not being able to address the issue at all. Sometimes in this sort of situation, instead of saying that the theory predicts the result, we say that the theory is consistent with the result. Sooner or later, better understanding of the theory and newly available input information lead to good predictions and tests. It would be nice if all tests were clear and conclusive, and eventually in physics they are. But with RIP, the more usual situation is the cloudier sort described in this section.

WHERE ARE THE SUPERPARTNERS?

The particles of the Standard Model include the electrons and quarks (explained in the next chapter) that we and all the objects in our world are made of. We'll see later that the main test of the validity of the idea that our world is supersymmetric is the existence of a set of previously unknown particles called *superpartners*. As I write, the superpartners have not yet been directly observed. Where are they? Why don't we see them? We think there are two parts to the answer. All but one of the superpartners are expected to be typically as heavy as the heavier of the Standard Model particles and therefore, like them, to decay (*decay* is explained in Chapter 2) rapidly into lighter particles. If nature is indeed supersymmetric, all the superpartners can be created in collisions at laboratories, as we will discuss in detail in Chapter 5, and they are also created about once every few minutes in collisions of cosmic rays at the top of the earth's atmosphere somewhere around the world. But they decay rapidly, and there is no way to detect them in the atmosphere. Only now are colliders and detectors at laboratories achieving the energies and sensitivities needed to detect the superpartners explicitly, at least if our thinking about their properties is more or less right.

We think that when superpartners decay, there has to be a lighter superpartner among the particles they decay into. They have to decay into lighter particles, so eventually each superpartner decays into Standard Model particles such as electrons or quarks or photons, plus the lightest superpartner. The lightest superpartner is expected to be stable because there is no lighter one into which it can decay. Then all of the lightest superpartners that have been created during the Big Bang, in collisions, and in decays of heavier superpartners since the Big Bang, should still be around and spread throughout the universe (except for a calculable small number that can annihilate on each other). This is the subject of Chapter 6. The estimates are that these relic lightest superpartners can make up a significant part of all the matter in the universe—a part called the dark matter—with about one superpartner in every grapefruit-sized region around us. That's where we think the superpartners are.

THE BOUNDARIES OF SCIENCE HAVE MOVED

One of the innovations in thinking that made modern science possible was focusing on how the world worked rather than on why the world was the way it was. Four centuries ago, "why" was left in the realms of religion and philosophy. "Why" questions were recaptured by science first in biology, when Darwin made them scientifically legitimate two and a half centuries after modern science began. It took over a century longer before physics began to deal with the "why" questions.

Before about the 1980s, the questions physics could address clearly had limits. Big questions such as where the laws of nature came from, and why there was a universe at all, were out of bounds. People could argue that each question answered would give rise to more, so there would always be voids in scientific knowledge, science would always have to take some fundamental aspects on faith, and some things were unknowable. What has changed is that now all of these big questions have become technical research questions. We don't know yet whether they will have scientific answers, but they are RIP. Most of the people working on them expect answers. Moreover, today a number of active researchers don't expect that there is anything necessarily unknowable concerning fundamental questions about the physical universe, nor do they expect that new questions about the ultimate laws of nature will necessarily continue to arise. We will return to these issues in Chapter 10. Supersymmetry does not itself provide the final answers to these big questions, but if our current ideas are right, it will provide the way to ensure that these questions can be studied as normal science and that proposed answers can be tested in normal ways. In the following chapters we will learn how to find out whether nature is indeed supersymmetric—and how the big questions can be studied if it is.

.... 2

The Standard Model
of Particle Physics

To better understand the reasons why we think supersymmetry will be discovered, and how it extends our present description of nature, we need to know a little about the main achievements of over two centuries of physics. Past study has led to the establishment of the Standard Model of particle physics, a complete description of the basic particles and forces that shape our world. In this chapter we will consider first the five forces and then the particles. The world we see is built entirely of three particles: the electron and two particles similar to the electron called quarks. There are more particles—antiparticles, neutrinos, more quarks and more particles like the electron, and Higgs bosons. We understand why some of the additional particles exist, but not others. Then we will learn a little about Feynman diagrams, a way of picturing how particles interact (for practitioners it is also a way to calculate the behavior of particles). Next we consider a property of all particles called spin and how it leads to categorizing particles into two groups, bosons and fermions. We will see later that at a deeper level, supersymmetry merges these two categories. Last we look at reasons why the Standard Model is not expected to be the final stage of particle physics, even though it successfully describes phenomena and experiments.

THE FORCES

When Gauguin painted "Where do we come from? What are we? Where are we going?" we knew only of the electrical, magnetic, and gravitational forces. There was controversy about whether atoms existed. The first particle to be discovered, the electron, had just been found. Radioactive decays of nuclei ("radioactivity") had just been noticed. These decays could not be explained by the known interactions, so physicists realized that another force was needed to describe nature's behavior. It was called the weak force, because its effects were rare and essentially never occurred when two interacting objects were separated by distances larger than an atom. In 1911 atoms were found to have a nucleus, and it was discovered that heavier nuclei have a number of protons in them. Because physicists knew that the repulsive electrical force pushed the protons apart, it became clear that yet another force—the nuclear force—must exist to bind the protons together into a stable nucleus. The effects of the nuclear force also can be felt only at tiny distances, no larger than a nucleus. Although Newton had correctly foreseen three centuries ago that there could be forces that had effects only at small distances, finding evidence for them took over two centuries.

These five forces (the electrical, magnetic, gravitational, weak, and nuclear forces) account for all we observe in nature. The interactions of particles with a "Higgs field" (more about this later, especially in Chapter 7) can be thought of as giving mass to the particles, so one can think of the Higgs interaction as an additional force. There may in a sense be other forces, but they are not relevant for the behavior of particles or for how particles combine to make up the world around us. (For completeness, let me mention two at this point. One is a force that recent evidence suggests causes the universe to expand more rapidly than it would if only the gravitational force affected its expansion. This force does not in any way affect the behavior of individual particles. For historical reasons it is called the cosmological constant. The other possible forces have effects only at *extremely* tiny distances far smaller than the size of a proton.) We will return to possible additional forces later briefly; the five known forces are the important ones for our purposes. Any others do not affect how our world works, though understanding them may be essential to achieve a complete picture of the laws of nature.

Ever since Charles-Augustin de Coulomb showed, over two hundred years ago, that the electrical and gravitational forces depend in the same way on the distance between interacting objects, and that therefore the formulas describing them have the same form, physicists have tried to unify our understanding of them. In the second half of the nineteenth century, using the work of Michael Faraday and others, James Clerk Maxwell succeeded in relating the electrical and magnetic forces, in the sense that electrical forces that vary in space or time generate magnetic forces, and vice versa. By the 1960s we knew of the five forces, the electrical and magnetic ones being unified into electromagnetism. There was no theory at all of the weak and nuclear forces, only a few known regularities of their behavior. By the 1980s, however, the Standard Model had emerged and had been well tested. It was a complete description of the weak, electromagnetic, and strong (nuclear) forces, fully consistent with quantum theory and special relativity. The progress over two decades was spectacular—in that short period, we went from a crude awareness of the weak and strong forces to their comprehensive description.

The picture of the electromagnetic force that emerges is that electrons, and any particles that have electric charge, interact by exchanging photons. The photons can carry energy between the electrons; two electrons can scatter off one another by exchanging a photon; and an electron and a proton bind by exchanging many photons, which provide an attractive force that keeps the electron and proton connected in a stable object, a hydrogen atom. All the forces work in a similar way. The gravitational force arises from the exchange of gravitons. The analogous particles for the weak interactions are called W and Z bosons, and for the strong force they are called gluons. In all these cases, we speak of the photon, W and Z bosons, and gravitons as "mediating" the forces. There is one more subtlety in connection with my use of the name *strong force* here. Because of the history, I cheated a little when I listed the nuclear force above. It turns out that the nuclear force between protons and neutrons that binds them into nuclei is not the fundamental force. Rather, the basic force is the strong force between quarks, and the nuclear force is a kind of residual effect after quarks are bound into protons and neutrons. We'll return to it after we describe the quarks.

MASS, DECAYS, AND QUANTA

Two frequently used words about particles that can be confusing are *mass* and *decays*. For our purposes, mass essentially means weight, and we don't need a more precise definition. Some particles, such as photons, don't have any mass. The masses of particles are measured mostly by bouncing particles off each other, since how much they bounce is related to how heavy they are.

Nearly all particles are unstable and decay into others. The word *decay* has a technical meaning in physics—one particle disappears, typically turning into two or three others. A major difference between the way *decay* is used in physics and its use in everyday life or biology is that the particles that characterize the final state are not in any sense already in the decaying particle. The initial particle really disappears, and the final particles appear. The photons that make up the light we see provide an example: The photons emitted from a light bulb when it is turned on are not particles that were in the bulb just waiting to come out, and photons that enter our eyes after bouncing off an object are absorbed by the molecules in our eyes and disappear. All particles can be created or absorbed in interactions with other particles. Most particles were created during the Big Bang, and particles can be created in collisions, e.g., at colliding-beam accelerators. Which particles can appear or disappear in any interaction is not arbitrary but is, rather, fully determined by the interactions described by the Standard Model (and its extensions such as supersymmetry).

Two constraints keep a few particles from decaying. First, the total amount of energy in any process must not change—we say energy is conserved. Imagine a decaying particle at rest; its energy is just its mass. That mass is divided into the sum of masses of the particles created in its decay, plus some energy of motion of those particles. Since energy can't appear from nothing, the particles created in the decay must be lighter than the decaying particle. Second, most particles carry some charge (electric charge and some similar charges associated with other forces). Charges are also conserved—the total amount of charge can't change in a process. Because of these two constraints, several of the lighter particles (electrons, for example) do not decay. Whenever basic particles are able to decay, the decay is typically rapid, occurring in a tiny fraction of a second.

Electromagnetic waves carry energy from antennas to our radios and from light bulbs to our eyes. One of the things quantum theory has taught us is that the energy is carried in little chunks (or quanta), photons. Any electric charge sets up an electric field around itself, and when that charge oscillates (say, in an antenna), it radiates the electromagnetic wave or the photons. We have also learned that there is a gravitational field associated with mass, and there are other fields associated with matter—with electrons and other particles. All the particles can be thought of as the quanta of the fields. In this book we won't make any technical use of such ideas, but every time we speak of some kind of field, we will associate quanta and therefore particles with it, and vice versa.

THE PARTICLES: DO WE REALLY KNOW THE FUNDAMENTAL CONSTITUENTS OF MATTER?

When viewed from the particle side, the Standard Model is remarkable. What are we made of? What happens if we keep cutting an apple, or a person, into smaller and smaller pieces? Do we reach a smallest piece? What could it be? What are the stars made of? Everyone has wondered about such questions. The answer is that everything we see in the universe, from the smallest cell to flowers to people to stars, is made of three kinds of matter particles, bound together by gluons and photons. When we get to big objects such as planets and stars, gravity binds too. The three matter particles are the familiar electron and two particles similar to the electron called quarks, the up quark and the down quark. *Up* and *down* do have a technical meaning, but it's not important here. All of the basic particles carry various amounts of charges, electric charge and weak charge and strong charge. The weak and strong charges are somewhat like electric charge, but because they have no effects outside atoms, they are not familiar to us in everyday life. The main difference between electrons and quarks is that the quarks carry the strong charge, so they interact via exchanging gluons, whereas electrons do not interact with gluons at all. Electrons and quarks all have electric charges and masses (weights) in different amounts. Sometimes electrons are denoted by e, up quarks by u, and down quarks by d.

How do these seeds form our world? The quarks bind together to make neutrons and protons, neutrons having an up quark and two down quarks, protons having two up quarks and one down quark. The quarks interact via the strong force, mediated by gluon exchange. The edges of protons and neutrons are not sharp—gluons range outward a little before they are brought back by the attractive force of the quarks. Gluons that are a little outside the proton and neutron in turn exert an attractive force felt by other protons and neutrons; this residual strong force is the nuclear force that binds the protons and neutrons into nuclei. The nuclear force is strong enough to hold together many nuclei, with from 1 to 92 protons and varying numbers of neutrons. The electrical repulsion felt by one proton (due to all the other protons in the nucleus) increases as the number of protons increases, and with more than 92 protons the nucleus becomes unstable. This is why there are only 92 naturally occurring chemical elements. The electromagnetic force, mediated by photons, binds electrons to nuclei to make atoms, the atoms of the chemical elements. Outside an atom there is a residual electromagnetic force (analogous to the residual strong force outside protons and neutrons described earlier in this paragraph) that binds atoms into molecules. Atoms and molecules build up rocks, cells, and all of the world around us. All of the marvelous complexity and color and structure of our world arises from these simple foundations.

Always in the past when objects were studied at smaller distances, they turned out to have structure. The atoms of the 92 chemical elements were not the atoms invented by the Greeks as the basic indestructible units of matter, because they turned out to have a nucleus surrounded by electrons. The nucleus turned out to be made of protons and neutrons. Protons and neutrons themselves turned out to be made of quarks bound by gluons. Why, then, do we think that the electrons and quarks are the true Greek "atoms" and that despite a history of smaller and smaller units, the progression stops with them? There are three kinds of reasons that lead most particle theorists to think we have finally reached the end of the line.

The first reason, significant but less important than the others, is that investigators have tried by many means to determine whether electrons, quarks, W bosons, and gluons show any evidence of structure, and they have not found any. These experiments have probed perhaps 10,000 times further than it took to see structure in the past, but electrons and quarks continue to behave as point-like objects with no parts.

The second reason is that always before, the prevailing theory (that of atoms or nuclei or protons) did not agree with experiment or even make sense unless these objects had structure at smaller distances, but the Standard Model is different. Because the Standard Model is a quantum theory, it is possible to ask how the forces behave at smaller and smaller distances and to do calculations to answer the question. Suppose that the basic particles were indeed point-like. How strong would the forces be if we could examine them not just down to almost 10^{-18} meter, where experiments can presently be done, but at a million million times smaller distances? The calculations show that in the Standard Model, all the interactions become weaker at smaller distances. The Standard Model does not lead to inconsistencies, no matter how small the distance that is probed. The Standard Model theory does not even have any possible way to include structure at smaller distances, no parameters for the size or for possible parts—it is a consistent theory without such parameters. The first reason, suggestive as it is, cannot be compelling because the next experiment may show evidence of structure (though it's not expected to). But having a theory that is confirmed by many experimental tests *and* can be extrapolated to much smaller distances constitutes a very powerful and convincing argument.

The third reason is also very compelling, and we will discuss it in more detail later in another context, but it is worth mentioning here. As stated above, we can calculate a prediction for the behavior of all the forces at smaller and smaller distances even though we cannot do the experiments. What emerges is that *if* the electron and quarks are structureless, then the different forces become more and more similar at smaller distances, and at a sufficiently small distance they become indistinguishable—a property that suggests how the two-centuries-old goal of unifying the forces might be realized. If electrons and quarks have structure, this does not happen. Unless this unification is a coincidence, it is a strong argument that no further structure exists.

When we say that the quarks, electrons, photons, and other bosons are the true fundamental particles, we do not mean they will not be reinterpreted, as vibrating strings, perhaps, or related objects instead of point-like objects. Maybe the particles will come to be thought of as analogous to the musical notes made by a vibrating violin string, with the string as the fundamental thing. But there will still be a vibration identified as an electron, another as an up quark, another as a photon, and so on. These basic constituents are not expected to be replaced by others. We will follow Lucretius, the Roman poet who wrote so presciently about science and nature two millennia ago, to "… reveal those atoms from which nature creates all things [and] call them the 'primary particles.'"

PARTICLES AND FIELDS

Newton's theory of gravity worked wonderfully to describe the moon and projectiles and the tides and much more, but it had a conceptual flaw. It seemed to work by some kind of instantaneous "action at a distance." How did the moon know at some instant to turn its path and stay near the earth instead of going straight off into space? Many people, especially philosophers, were upset about that and resisted accepting the gravity theory. Physicists are used to thinking about *how* something works separately from *why* it works that way, so they went ahead and developed and applied Newton's theory. It took over two centuries to fully understand what was actually happening.

The initial development wasn't intended to solve the problem but turned out to be the first of several major steps that did solve it. In order to simplify solar system calculations, where the effects of all the planets on each other had to be included, people developed a procedure different from Newton's but equivalent. Instead of writing the forces exerted on a planet by all the other planets and the sun, which was very complicated to do, they considered one planet at a time as though it were the only object in the universe and imagined that it set up a gravitational "field" through all space—a field defined such that if another planet were put at a given point, the resulting force between the two would be exactly the Newtonian one. At a given point in space, one could add the fields from different bodies and obtain the total field, which was then what would be felt by another planet if one were there. Originally, this was viewed as just a trick to simplify calculations.

It was Michael Faraday who largely convinced physicists to think of fields as real physical things rather than as calculational devices. He did that in the context of electric and magnetic fields. Most of us have seen what happens when iron filings are put on a piece of paper over a magnet—they align themselves into a visual representation of the magnetic field. We now think of electric and magnetic and gravitational fields as being as real as rocks and people. Normally we are not aware of those fields, but you can confirm that they are all around you by turning on a radio.

The next stage came when Einstein demonstrated that no signal or bit of information could move faster than the speed of light and, in particular, that electromagnetic and gravitational fields spread out at exactly the speed of light. That's very fast, but not instantaneous. For classical (pre-quantum) physics, the action-at-a-distance problem was solved. The final stage came in the context of quantum theory. First photons were interpreted as the quanta of the electromagnetic fields. Then it was realized that electrons and all particles should be viewed as indivisible quanta, each of their own fields. The effects of the fields are transmitted by the quanta. Today we think of the fields as the basic ingredients of the universe, and we often speak of the particles and the fields interchangeably.

The modern way to describe nature at the level of the particles is for theorists to write down an expression called a Lagrangian (after Joseph Louis Lagrange, who first formulated Newton's theory in this way). The Lagrangian contains the symbols representing the quantum fields for all fundamental objects (electrons and quarks, photons and gluons, and W and Z bosons) and *only* for the fundamental objects. All composite objects (anything with structure, such as protons, nuclei, atoms, and so on) are built up from the particles described by the Lagrangian. Given the Lagrangian, the rules of quantum theory tell how to calculate the behavior of all systems in principle, though in practice some are so complicated that the calculation can't be actually carried out. Knowing the Lagrangian means knowing the fundamental particles and forces. The achievement of the Standard Model was to discover the Lagrangian of the world that we see.

There Are More Particles

I oversimplified a little in the first part of this chapter. Although it is true that everything we *see* is made of the electron, up quarks, and down quarks, research has uncovered the existence of more particles. We do not see them, and they do not directly show up in our world, but we can do experiments to reveal them. We understand why some exist, but others are still a mystery. A brief look at some of these additional particles follows: antiparticles, neutrinos, more quarks and leptons, and Higgs bosons.

- ANTIPARTICLES. In 1928 Paul Dirac combined quantum theory and Einstein's special relativity in the description of the electromagnetic interaction. In the process he showed that the resulting equations had a surprising property. We have seen that particles are represented by the solutions of the basic equations. Dirac showed that if the behavior of one particle—say, an electron—was described as a solution of the basic equations, then the equations must also have another solution with the same mass as that particle, but with all charges opposite in sign. In the case of the electron, which has a negative electric charge, the new particle would have a positive electric charge. If one solution exists in the real world, so must the other. This is somewhat like saying that the square root of 4 is both +2 and −2, because either of them squared gives 4. To make an equation we could write $x^2 = 4$, with solutions $x = +2$ and

$x = -2$. If one is a solution, so is the other. Similarly, if one particle exists, so must the other. The two particles are called *antiparticles* of each other. Dirac's result holds for all quanta or particles. If a particle has no charges, such as a photon, it is its own antiparticle—it and its antiparticle are not distinguishable.

This was the first major example where a previously unknown particle was predicted by theoretical arguments, and it did not add just one particle but doubled the number of particles: an antiparticle for each particle. What is amazing is not the existence of "antimatter" but that a purely theoretical argument predicted the existence of unknown particles and later they were actually found. Human reasoning and guidance by a theory combined to reveal a previously hidden part of nature. We will see that supersymmetry predicts another doubling of the number of particles.

Although there is a certain popular mystique about antimatter, it is not justified. The particles and their antiparticles are all just particles with different charges. The antiparticles have all been observed. Which is called the particle and which is called the antiparticle is just a convention. Antiparticles are individually as permanent as the associated particles, and they behave the same way the particles do (once the opposite sign of the charge is taken into account). For example, antielectrons (often called positrons), and antiprotons can bind to make antihydrogen. But because electrons and positrons have opposite charges, their net charge is zero, and so a pair of them can annihilate, if they collide, into photons that carry off energy. Indeed, anytime a positron finds itself in our world, it soon hits an electron and annihilates. Thus there are normally few positrons around. Similarly, antiprotons annihilate with protons, so few antiprotons are around. We will return later to consider why our world is mainly matter and not antimatter.

- NEUTRINOS. In 1930 experiments seemed to suggest that some nuclear decays did not conserve energy—less energy was detected in the final state than in the initial one. Wolfgang Pauli proposed that the energy was being carried off by an unseen particle, one that had no electric charge and interacted only via the weak interaction, in which case it would pass right through normal detectors without any effect. Yet it would carry off energy. If that were true, certain predictions could be made for the motion of the other particles that emerged in the decay, and those predictions were verified, so within a few years most experts were convinced from the indirect evidence that "neutrinos" indeed existed. It was not until 1958, however, almost 30 years after they were postulated, that they were directly detected. Neutrinos are often denoted by the Greek letter nu (v).

- MORE QUARKS AND LEPTONS. There are two more particles like the electron. All their properties are identical to those of the electron, except that they are heavier, and each is associated with its own neutrino. Although we don't have any idea why that is so, the Standard Model theory still correctly describes the behavior of these particles. In experiments, various numbers of these particles are produced going in various directions. They are unstable particles, decaying in less than a millionth of a second into electrons, positrons, neutrinos, and photons, so they do not stay around after they are produced, nor do they end up in anything we see. In all cases, the Standard Model correctly predicts the number of them, their directions, energies, and lifetimes, and all other aspects of their behavior. One is called the muon, denoted by the Greek letter mu (μ), and the other is known by the Greek letter tau (τ). Because they behave in very similar ways, the electron, the muon, the tau, and their associated neutrinos are grouped together in a class of particles called leptons.

There are also two more quarks just like the up quark except that they are heavier. Here again, the Standard Model correctly predicts all aspects of their behavior. They are called the charmed quark (denoted c) and the top quark (t). And there are two quarks just like the down quark except that they are heavier. They are called the strange quark (s) and the bottom quark (b). Although the Standard Model can fully describe the behavior of all of these extra quarks and leptons, it cannot tell us why they are there.

Usually these quarks and leptons are grouped into three so-called families. This apparent replication of particles into three groups, each group exactly identical to the others except for the increasingly heavy masses of the particles that make up each group, is one of the most surprising and disconcerting discoveries of the past century in particle physics. It is called the "family problem." The first family consists of the electron and the up and down quarks that together make up what we see, plus the neutrino associated with the electron. The second family includes one of the particles like the electron but heavier (the muon), the neutrino associated with the muon, and quarks that are like the up and down quarks but heavier (the charmed and strange quarks). The third family has an even heavier electron-like particle (the tau), the neutrino associated with the tau, and the other heavy quarks (top, like up, and bottom, like down).

There is good experimental evidence that there are no more such families of quarks and leptons. We have no idea why nature has three families instead of just one. As far as we know today, one family would have been sufficient to build our world. It is not that we have not yet been clever enough to figure out from the Standard Model why there are three families—the Standard Model simply does not contain principles that could explain this. We hope that a future extension of the Standard Model will; an attractive feature of string theories is that they appear to be able to address this issue, although we do not yet know whether they do so correctly.

- HIGGS BOSON(S). The Standard Model predicts another kind of particle should exist, a "Higgs boson." Experimental proof of its existence will complete the Standard Model. The precise properties of the Higgs boson, and how many Higgs bosons exist, will give us clues to how the Standard Model will be extended; for example, supersymmetry predicts that at least five types of Higgs bosons exist. The existence and properties of Higgs bosons are a special test of the Standard Model because Higgs bosons are different from any previously known kind of particle. Leptons and quarks are basically all particles like electrons, that merely carry different electric, weak, or strong charges. Similarly, all the quanta that mediate the weak and strong interactions are like the photon. However, the Higgs bosons that the Standard Model predicts are really a new kind of matter. If they are found, their prediction and discovery will stand as a spectacular achievement of human reasoning.

We will return to consider Higgs bosons in more detail in Chapter 7. At present (late 1999) there is significant indirect evidence that a Higgs boson exists with a mass about right for what is predicted by supersymmetry. Experiments in 2000 could detect direct evidence for a Higgs boson at the CERN collider LEP in Geneva, Switzerland. (See the list of abbreviations in the Glossary at the back of this book. These facilities are described later in the book.) The upgraded collider at the Fermi National Accelerator Laboratory west of Chicago may produce and detect Higgs bosons if the facility runs at sufficient intensity, and if the detectors perform sufficiently well, when it starts to take data hopefully in 2000 or 2001. To collect enough Higgs bosons to see a signal at Fermilab should take 3 to 5 years, depending both on details of the properties of the Higgs boson and on how effectively the collider and the detectors perform. Fermilab will be able to find Higgs bosons—if they exist—with the properties supersymmetry implies they should have. The Large Hadron Collider under construction at CERN and scheduled to begin operating in 2005 will produce large numbers of Higgs bosons (again, assuming they exist).

NEW IDEAS AND REMARKABLE PREDICTIONS
OF THE STANDARD MODEL

The Standard Model is based on two major theoretical principles. The first is that *the underlying laws of nature do not change when certain apparently different particles are interchanged in the equations.* We say that the theory is invariant under those interchanges; such an invariance is an example of a symmetry. One such pair is the electron and its neutrino. Recall that these two have similar properties and behavior but differ in that the electron has electric charge, whereas the neutrino does not, and in that they have different mass. Imagine a kind of abstract space (not our real space) where particles can be represented by arrows. If a particular arrow points one way, it is an electron; if it points the opposite way, it is the electron neutrino. The charge is a label that tells which way the arrow is pointing at the moment. Particles don't change the direction they point when they are sitting around, but they can when they interact. Other possible interchanges include up quark for down quark (that is, interchange the symbols for the up quark and the down quark in the equations). Requiring that the equations not change when the particles are interchanged imposes powerful constraints on the form of the theory and on its predictions. One clear prediction is that if one particle that should belong to a hypothetical interchangeable pair is found, then the other must also exist. The charmed quark and the top quark were both predicted to exist by this argument, and then found with the expected properties.

Another prediction based on this argument is that the particles must have the same mass if they can be interchanged, since the equations can depend on their mass. Some technical arguments show that this is possible only if all of the particles (quarks and leptons and W and Z bosons) have zero mass. Unfortunately, that does not agree with the data. These particles do have mass, and this is where Higgs boson physics enters. It is the interaction with the Higgs boson that allows all particles not only to have mass but to have different masses. That interaction has a special form that makes it possible for the equations to retain their invariance under interchange of the particles but still enables the particles to have (different) mass. Photons and gravitons and gluons don't directly interact with Higgs bosons, and therefore they remain massless. The existence of the Higgs physics is essential to a consistent theory that can describe the actual particles.

The symmetry of the Standard Model under interchange of certain particles is well hidden—it showed up only after physicists guessed it was there and then used it to predict some implications that could be tested experimentally. Once this symmetry under interchange was indeed demonstrated for the Standard Model, physicists began to take more seriously the possibility of other well-hidden symmetries in nature. That made it easier to think about having a symmetry under interchange of bosons and fermions—that is, supersymmetry (bosons and fermions are defined explicitly later in this chapter).

The second principle is really a consequence of quantum theory rather than a new idea, but it took several decades before it was fully understood, and it really emerged only as part of the Standard Model. *If there is a particle, such as an electron, carrying a charge, then it is impossible to make a consistent quantum theory unless an additional field exists and interacts with that particle.* This additional field has precisely the properties of the electromagnetic field, so it can be interpreted as being the electromagnetic field. Since the quanta of the electromagnetic field are photons, the photon must exist once electrons do, given that quantum theory provides the rules by which nature operates. Thus in the Standard Model the photon is not an extra or separate part of the world, with electrons and photons happening to interact—rather, once the electron exists, so must the photon. The existence of the photon is explained. With the Standard Model we finally understand what light is.

The same argument applies to the weak and strong charges. Once any particle carries a weak or a strong charge, there must exist particles like the photon but associated with the weak and strong forces. These particles are the W and Z bosons for the weak force, and the gluons for the strong force. The existence of these particles was predicted by this argument, and they were found, in planned experiments, with exactly the properties they were expected to have. Its having correctly predicted the existence of these particles is one of the strongest reasons why we are confident the Standard Model is here to stay.

EXPERIMENTAL FOUNDATIONS
OF THE STANDARD MODEL

A mere description of the Standard Model itself does not do justice to the complex and interesting experimental history and tests that both led to its formulation and gave us confidence that it indeed describes nature correctly. Without data of many kinds—the Lamb shift of hydrogen energy levels, inelastic scattering of electrons and neutrinos, parity violation, neutrino beams, pion decay, neutral currents, W and Z boson production and decay, charmed quark production, and much more—the Standard Model would not have been developed. It is too far removed from our world to have emerged from theory alone. These and others were major crucial experiments, though we will not consider them further in this book. Doing those experiments required new developments in accelerator and detector physics and in computer technology, the insights to carry out the right experiments, and the skills to get them right. Later we will discuss some of the experiments that might be performed to establish and test supersymmetry and even string theory and will get a better sense of the role of experimentation. Several of the books listed in the annotated bibliography ("Some Recommended Reading") at the end of this book describe the experimental foundations of the Standard Model.

PICTURING STANDARD MODEL PROCESSES:
FEYNMAN DIAGRAMS

All particles do is interact with other particles. Before and after interacting, they travel about freely. Calculating how likely various interactions are, and the details of what happens if the interactions occur, can be extremely technical and time-consuming. In the 1940s Richard Feynman invented an ingenious way to greatly simplify calculations. He developed a way to draw a few diagrams so that they both become an algorithm for doing the calculations and also enable one to picture how and how often the processes of interest will occur. The algorithm was far simpler than what one had to use before Feynman created the diagrams; the increased simplicity was very important because it allowed many more calculations to be carried out that could then be compared with experiment to test the theory and develop it. Perhaps even more important, with Feynman diagrams it became possible to think about and explain how particles behave, in ways that anyone could do on blackboards or scraps of paper. It is hard to overemphasize the importance of innovations, such as Feynman diagrams, that help us think clearly. In this section I will outline some of the features of Feynman diagrams that we can use in this book to visualize what particles do. Here we will focus on the Standard Model; later we will add the additional interactions introduced by supersymmetry.

Most of the basic interactions from which processes are built can be represented as pictures where three particles touch at a spacetime point. Arrows can be put on the lines to tell whether the particles are entering or leaving the interaction. Suppose we want to show an electron moving along and absorbing a photon. It is customary to label the photon by the Greek letter gamma (γ) and the initial and final electrons by e and e'; the prime on the final electron shows that while absorbing the photon, it could gain or lose energy and momentum and change direction. We would draw this as shown in Figure 2.1.

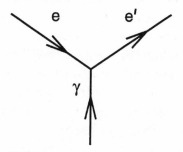

FIGURE 2.1.

An electron could also turn into its partner the electron neutrino (ν_e) by absorbing one of the W bosons that mediate the weak force. We would draw that as shown in Figure 2.2.

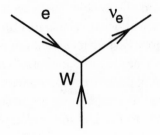

FIGURE 2.2.

With such figures, by appropriately choosing the particle types (electron or other leptons, up or down or other quarks, photons, W or Z bosons, or gluons) and arrow directions, one can write all of the allowed vertices in the Standard Model. Some of the allowed ones are shown in Figures 2.1 and 2.2; others are $u + W \rightarrow d$, $Z \rightarrow e + e'$, and $u + g \rightarrow u'$. It may seem as though we are just writing all possible interactions, but many are not allowed by the Standard Model. Any process that does not conserve the amount of charge is forbidden (for instance, $e + \gamma \rightarrow \gamma'$ has charge in the initial state, but not in the final state so it cannot occur). In addition, a number of processes that would otherwise be allowed are predicted by the Standard Model not to occur in nature (and they don't). Examples include $\gamma + \gamma \rightarrow Z$ and $e + \gamma \rightarrow \mu$, and there are many more.

Feynman then gave the rules: (A) Join the allowed Standard Model vertices into all possible connected diagrams that have the initial and final states you are interested in. This shows all the interactions that can join these states. (B) To find the probability for each process actually to occur in nature, follow a certain procedure that leads to a calculation of the probabilities. For any theory where the interactions are not too strong, which fortunately includes any process with very energetic beams in the Standard Model, or any decay of a heavy particle, the more vertices that occur in the diagram, the less probable it becomes, so usually the simplest few are enough to calculate the result. We will not calculate any of them or need to know any more about them, but we will occasionally show what happens by using the simple diagrams. Although Feynman's procedure may seem rather *ad hoc* or random, it is not. He and others were able to show that drawing the diagrams completely reproduced the predictions of the full mathematical theory—and did so in a far simpler way than was available before. Graduate students are taught a systematic procedure for constructing the diagrams and their implications. Remember that a diagram not only is valuable as a picture of what happens but also implies a set of rules with which the physicist can calculate in detail what happens. It is truly remarkable that following this procedure describes all observed phenomena in our world: the building up of nuclei, atoms, molecules, organisms, planets, and stars from the basic constituents, and the behavior of all these systems.

Let's look at two examples of processes that can be obtained by combining vertices. We know that particles of like electric charge repel. If two electrons are heading toward each other, they repel, veer off, and separate, exchanging a photon. That is shown in Figure 2.3 (below).

This is not meant to be a literal picture of what happens; the electrons are not shown actually moving toward each other. It is a guide to thinking. One can imagine time moving from left to right in the picture. The electrons move along, exchange the photon (thereby interacting), and separate. One electron emits the photon and the other absorbs it. This is how the electromagnetic force works.

All the other forces work similarly, via the exchange of photons, or W and Z bosons, or gluons, or gravitons. We can see here a hint of an idea that is not really a part of the Standard Model but that suggests its extension toward supersymmetry and even string theory, as we will see later in the book. All forces are generated by exchange of particles, so once there are particles in the world, plus interactions such as the vertices of Figures 2.1 and 2.2 that allow particles to be emitted or absorbed, the existence of combined diagrams such as Figure 2.3 follows automatically. Therefore, the existence of forces (represented by Figure 2.3) is seen to be a consequence and not a new, independent effect.

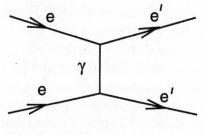

FIGURE 2.3.

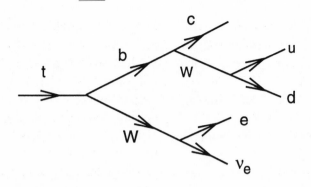

FIGURE 2.4.

Finally, consider one way in which a top quark (t) can decay (see Figure 2.4). The quarks (c, u, and d) and the positron (shown as e) enter the detector, and the neutrino (ν_e) escapes, carrying energy away. The W's and the b quark mediate the interactions, like the photon in Figure 2.3.

Later in the book we will examine some processes that we think will involve the particles predicted by supersymmetry, and we will use Feynman diagrams to illustrate interesting ones. Just read them along the arrows to understand the process. I hope they will make it easier to visualize what happens in processes wherein superpartners are produced and decay, as we consider some of those processes later in the book. If the diagrams aren't helpful to you, just ignore them.

Spin, Fermions, and Bosons

All particles have a property called spin. As with other features of the Standard Model, the term *spin* is suggestive but not precise—it is best thought of as an analogy. We don't yet know what spin is or why particles have spin, but we can fully describe the effects of having spin and every way in which spin affects the behavior of particles. Spin is another quantum kind of property. Particles can have amounts of spin only in certain chunks of a basic unit. The allowed values of spin are zero and half-integer and integer multiples of the basic unit: $1/2$, 1, etc.—the value of the unit doesn't matter for us. Particle spins are quantized in another way too. Their spin can point only in certain orientations. For a spin-$1/2$ particle, there are two orientations: One can think of an arrow pointing up or down along some direction (what direction doesn't matter).

The way quantum theory works leads to particles being divided into two classes, called bosons and fermions. The reason is a little technical but not hard to understand. First one has to be aware that all electrons are identical to each other. All electrons everywhere are indistinguishable. All photons are likewise identical to each other, and every one of each kind of particle is indistinguishable from others of that kind. This is not a puzzle; all electrons are quanta of the electron field, and we would expect all quanta of a given kind of field to be identical. Being identical to the others doesn't matter much if we are considering only one of them, but when more are involved there are some surprising implications in quantum theory.

The way nature works is that the descriptions of all particles with half-integer spins ($^1/_2$, $^3/_2$, ...) must change sign under interchange of identical particles, and the description of all particles with integer spins (0, 1, ...) must not change sign under interchange of identical particles. The former are called fermions, the latter bosons. Quarks and leptons are fermions. Photons, gluons, W, and Z (the particles that mediate the forces) are bosons.

For an individual particle, it doesn't matter much whether it's a fermion or a boson. However, when lots of particles get involved, the behavior of fermions and that of bosons are remarkably different. The existence of fermions is responsible for the stability of matter, whereas all of the force-mediating particles are bosons. Thus physicists normally think of fermions and bosons as very different kinds of particles. It was a great surprise to discover that there could be a symmetry that can relate them—supersymmetry.

BEYOND THE STANDARD MODEL

In a sense, the Standard Model has achieved the goals that science has traditionally set. Finally, for the first time in history, we have a complete description of how our physical world works. There are no puzzles or conflicts between theory and experiment. It is the underlying physical theory for all of chemistry, the behavior of matter, the behavior of stars, and all that we see. As I explained earlier, that does not in any way imply that doing science has ended, because the equations still need to be solved for many systems, and unexpected properties emerge for complex systems (see Chapter 3). Once we had the Standard Model, we could ask new questions about the universe as a whole, and we could inquire why the Standard Model took one form and not another. Many questions that were philosophical or entirely speculative have recently become normal research topics in physics, astrophysics, and cosmology. In this section I will list some of those topics. My purpose is not to give a detailed understanding, but to provide an overview of why so many theorists are confident that new physics that extends the Standard Model (such as supersymmetry) is just around the corner.

- We saw earlier that in order to describe the masses of the particles, it was necessary for the Standard Model to contain some new interaction that modified the particle interchange symmetry of the Standard Model in a very specific way, and that the Higgs physics could indeed do that. The Standard Model itself cannot provide a mechanism that produces or explains the Higgs physics in the required way, so some new physics beyond the Standard Model is essential. Supersymmetry can provide that new physics, and at present it is the only known way to do it.

- I mentioned above that if we effectively looked at the weak, electromagnetic, and strong forces at smaller distances by using the quantum theory to magnify them, we found that they looked increasingly similar to each other, although there was no apparent reason why that should happen. In the supersymmetric extension of the Standard Model, the forces essentially become identical at sufficiently small distances, a result that strongly encourages optimism about the possibility of truly unifying them.

- Planets farther out in the solar system take longer to orbit the sun, in accordance with Kepler's laws, which were discovered four centuries ago. When we examine the motion of stars in our galaxy, or of galaxies in clusters of galaxies, we find that Kepler's laws do not seem to apply if they are written to include only the visible stars, but they do seem to apply if a large amount of nonvisible matter ("dark" matter) is included. If we couldn't see the moon, we could deduce from the tides that it was there, once we understood how gravity works. Similarly, we can deduce that the dark matter is there. When this was done on all scales in the universe, we learned that far more dark matter is present than could be accounted for by Standard Model particles. Many possibilities for the dark matter have been suggested—that it is the lightest of the new superpartners predicted by supersymmetry is the favored one. Since we don't yet know which superpartner it is, it is often called the LSP, for lightest superpartner. We will return to this question in Chapter 6 and consider how we can learn experimentally what actually makes up the dark matter.

• People and planets and stars are made of protons and neutrons, but not the antiparticles of protons and neutrons. The Standard Model cannot provide a mechanism to explain how to start with a universe that initially is symmetric between particles and antiparticles but evolves to give the observed asymmetry. Note that this is a case where we do not know whether we can find an explanation—we could simply say that whoever started the universe set it off with an excess of matter over antimatter, but that is philosophically unacceptable to most physicists. One can imagine understanding a universe that began with equal numbers of particles and antiparticles, with no net number of either, but not a universe that somehow had more particles than antiparticles from the beginning. In fact, if certain conditions hold, it is possible to explain how such an asymmetry could evolve from initial symmetry as the universe aged. Those conditions can indeed hold if the Standard Model is extended in certain ways. There are actually several alternatives, and most of them rely on supersymmetry. We will return to this issue in Chapter 8.

• The Standard Model is a consistent theory whether or not neutrinos have mass. Technically, neutrinos with mass could be described in the Standard Model the same way as electrons, using the Higgs interaction. But experiments show that neutrinos have at most about a million times less mass than electrons, and it would not make sense for their mass to arise in the same way as the electron mass and yet be much smaller. Recently, experimental evidence has emerged that at least one kind of neutrino has a nonzero but very small mass. Since the late 1970s, theorists have been inventing ideas that would lead to theories in which neutrinos have mass. Since the mid-1980s, all experts in this area have agreed that it is extremely unlikely that neutrinos are massless and that all approaches to incorporating a mass for them into the theory require some new physics beyond the Standard Model. Here supersymmetry is helpful in formulating ideas but has not played an essential role so far.

- The phenomena listed so far (the need for a mechanism to account for Higgs physics, the matter-antimatter asymmetry, neutrino masses, and dark matter) are based on real data that we can hope to understand with new principles—with an extended theory that includes the Standard Model. We can also list a number of "why" questions that we can hope to understand. Why are the basic particles quarks and leptons rather than something else? What *are* quarks and leptons? Why is the Standard Model theory unchanged if we interchange electrons and their neutrinos, or up and down quarks, but not if we interchange (say) electrons and muons, or electrons and up quarks? To put it more generally, we can ask why the Standard Model is what it is. There were other reasonable and elegant theories that addressed the same issues as the Standard Model, but data showed the Standard Model to be the correct one, the one that actually described nature. Why are there three families of quarks and leptons? Why is it so hard to make a quantum theory of gravity? Do the different forces in fact become indistinguishable and somehow turn into one force at very small distances, so that we can have a simple picture of the universe as governed by one underlying force that just appears as several different forces when we look at it at our large distances? We could list more of these kinds of questions—this list suffices to make the point. We have no guarantee that we can ever answer such questions. But we can try, and in fact research so far has always suggested possible answers—sometimes in the context of superstring theory (Chapter 9). In every case, supersymmetry is relevant to the answers. We will discuss some of these possible answers in the following chapters.

Readers unfamiliar with particle physics may have been amused by, or had some other reaction to, the largely metaphorical terminology they encountered here: up and charmed and top quarks, gluons that bind, families, strings, and so on. Of course, all of these terms have precise meanings and correspond to specific quantities in equations. But we need to talk about them even though they do not correspond to things in our everyday world. Particle physicists have chosen to humanize them with familiar names that usually carry a reminder of the role of the particle or interaction they refer to. The names often help us to think metaphorically about the way the world works at tiny distances. Sometimes people misinterpret the apparently cavalier terminology as an irreverent one, but in fact it shows great respect for these new properties of the world that we have discovered as we have probed nature in new ways.

Readers who would like more information about the Standard Model can turn to my book *The Particle Garden* (see "Some Recommended Reading"). It describes the Standard Model in more detail, but simply enough for any curious reader, and it assumes no prior knowledge or technical background. It also includes chapters on the history of particle physics, experimental facilities and the experimental foundations of the field, and connections to astroparticle physics and cosmology.

An interesting perspective on physics beyond the domain of the Standard Model is provided by asking about the relative status of experiment and theory as a field progresses. Particle physics began as a field about a century ago; one could date it to the discovery of the electron, or perhaps a few years later to the use of particle beams, which allowed us to learn the atom had a nucleus. From then until the discovery of the Standard Model in the early 1970s, experiment was ahead of theory—there were many unexpected discoveries and essentially no successful predictions. Once the Standard Model was written, it explained many puzzles and made many dramatic and successful predictions. There were no experimental surprises for over two decades—all results either had been clearly predicted or at least were anticipated. Once we go beyond the domain of the Standard Model, the situation is mixed. Today theory and experiment are again making progress together, as they did throughout most of the history of physics. But the situation is different in an important way from that of the century before the Standard Model, because now the Standard Model provides a framework that allows new kinds of questions to be formulated.

Why Physics Is the
Easiest Science — Effective Theories

If we had to understand the whole physical universe at once in order to understand any part of it, we would never have made any progress. Suppose that the properties of atoms depended on their history, or on whether they were in stars or people or labs—we might not understand them yet. In many areas of biology and ecology and other fields, systems are influenced by many factors, and of course the behavior of people is dominated by our interactions with others. In these areas progress comes more slowly. The physical world, on the other hand, can be studied in segments that hardly affect one another, as I will explain in this chapter. If our goal is learning how things work, a segmented approach is very fruitful. That is one of several reasons why physics began earlier in history than other sciences, and why it has made considerable progress: It really is the easiest science. (Other reasons include the relative ease with which experimenters can change one quantity at a time, holding others fixed; the relative ease of improving experiments when the implications are unclear; and the high likelihood that results are described by simple mathematics, a property that enables investigators to deduce testable predictions.)

Once we understand all the segments, we can connect and unify them, a unification based on real understanding of how the parts of the world behave rather than on philosophical speculation. The history of physics could be written as a process of tackling the separate areas once the technology and available understanding allow them to be studied, followed by the continual unification of segments into a larger whole. Today one can argue that in physics, we are finally working at the boundaries of this process, where research focuses on unifying all of the interactions and particles—this is an exciting time intellectually. In recent years physicists have understood this approach better and have made it more explicit and formal. The jargon for the modern way of thinking of theories and their relations is the method of "effective theories." I'll use this bit of terminology in this chapter and occasionally later in the book. Sometimes this approach has been called a reductionist one. But the word *reductionist* has different meanings and implications for different people, so I won't use it here. For physicists, *reductionist* implies simultaneously separating areas to study them *and* integrating them as they become understood.

ORGANIZING EFFECTIVE THEORIES
BY DISTANCE SCALES

Probably the best way to organize effective theories is in terms of the typical size of structures studied by a particular effective theory, which we speak of as the distance scale of normal phenomena described by that theory. Imagine starting by thinking about the universe at very large distances, so large that our sun and all stars look like small objects from such distances. This is the effective theory of cosmology, where stars cluster in galaxies because of their gravitational attraction, and galaxies are attracted to each other and form clusters of galaxies. Because of gravitational attraction, everything is moving on a background of the expanding universe. The only force that matters is gravity. We can use simple Newtonian rules to describe motion—deviations due to effects described by quantum theory are tiny and can be ignored. We can study how stars and planets and galaxies form, their typical sizes, how they distribute themselves around the universe, and so on. It doesn't matter whether the particles that make up stars and planets are composed of quarks or not, nor does it matter how many forces there are at the small distances inside a nucleus. The large-scale universe is insensitive to what its contents are except for their mass and energy. Because of this indifference, cosmology can make progress regardless of whether we understand how stars work, whether protons are made of quarks, and so on. We can learn from astronomy data that there is dark matter. At the same time, if the dark matter is composed of particles, we cannot learn from astronomy or cosmology what kind of particles they are, because cosmology is largely insensitive to the properties that distinguish one particle from another, such as their masses and what charges they carry.

Next consider smaller distances, about the size of stars. We can study how stars form, how they get their energy supply, how long they will shine—that is, we can work out an effective theory of stars. While doing that, we can ignore whether the stars are in galaxies and whether there are top quarks or people. Let's go to a smaller distance, say people size, and consider the physics. Gravity keeps us on the planet, but otherwise it is the electromagnetic force that matters. All of our senses come from mechanical and chemical effects based on the electromagnetic force. Sight consists of photons interacting with electrons in our eyes, followed by electrical signals traveling to our brains. Touch begins with pressure affecting cells in the skin, leading to electrical signals propagating to the brain. Hearing starts with air molecules hitting molecules in the eardrum, interacting via electromagnetic forces. Friction, essential for us to stay in place or to move, is due to electromagnetic forces between atoms. We don't need to know about the weak or strong forces, or about the galaxy, to study people-sized physics. The energy that the sun supplies to the earth provides all of our food and essentially all of our energy, mostly from stored solar energy, but otherwise how the sun works does not matter. Here is a case where phenomena from one effective theory provide input to another in a very specific way—the earth can be viewed as a closed system except for the input of solar energy. From the point of view of the effective theory of the planet earth, how the sun generates its energy is irrelevant.

Now consider atomic size. Here we'll be able to see even better how powerful the effective theory idea is. To be able to use the basic equations that govern atoms, we have to input some information, essentially a few properties of the electron (its mass and electric charge and spin) and the same properties for each of the naturally occurring nuclei, if we want to describe the whole periodic table of the elements. A description of atoms doesn't need to take account of whether stars or galaxies or people exist, whether the nucleus is made of protons and neutrons, or whether protons are made of quarks.

Before we go to even smaller distances, this is a good place to stop and consider some of the implications of this way of thinking. When we have a tentative theory of atoms, we want to test its predictions and determine whether it explains phenomena we already know. The predictions for the lifetimes of atoms, for the energies of photons emitted by atoms, for the sizes of atoms, and so on depend on pieces of input information—on the masses and charges and spins of the electron and the nuclei. Every effective theory has some input parameters such as these. Without input information about the electron and nuclei, we could solve the equations but not evaluate the results numerically, so we could not test the theory, make any useful predictions, or tell whether we had the right theory. For example, the radius of the hydrogen atom is $h^2/m_e e^2$ (h is Planck's constant, m_e is the electron's mass, and e is the magnitude of the electric charge of the electron). If we measure the size of the hydrogen atom, we still can't check whether the theory is working unless we know (by measurement or calculation) the mass and charge of the electron and Planck's constant (all three were measured nearly a century ago).

Now we can emphasize an important aspect of effective theories. When we study the effective theory of the nucleus, we want to be able to calculate the masses and spins and charges of the nuclei—those parameters are results derived in the effective theory at that level. But for the next higher level, the atom, those parameters are the input. They can be input if they are measured, whether or not they are understood in the effective theory of nuclei. Similarly, the electron is a fundamental particle whose mass and electric charge will, it is hoped, be calculable someday in a fundamental theory such as string theory, but for the effective theory of atomic physics it does not matter whether they are calculable or understood, if they have been measured. We have known the numerical value of the mass of the electron since the beginning of the twentieth century, but we do not understand why it has that value. The value of the electron's mass can be input into every effective theory that depends on the electron. Every effective theory so far has some input that is for it a "given," not something to be questioned; for the effective theory of atoms, the mass of the electron is such a given.

If a theory has inputs, it is an effective theory. From this point of view, the goal of particle physics is to learn the ultimate theory at the smallest distances, recognizing it as the theory for which no parameters have to be input to calculate its predictions. For the ultimate theory, it is not satisfactory to input the electron mass; rather, it is necessary to be able to calculate that mass from basic principles and to explain why it has the value it does.

Every effective theory is based on others: It is effective theories all the way down, until the ultimate one. Each effective theory has certain structures that bind together at its level—stars, atoms, nuclei, protons. As viewed from the effective theory of stars, stars are made of nuclei and electrons bound by the gravitational force, but for the effective theory of cosmology, they are just inputs characterized by a mass and brightness. For the effective theory of nuclei, nuclei are bound states of neutrons and protons, but for the effective theory of atoms, they are merely point-like inputs. All systems and structure are inputs at one level of effective theories but are something to be derived and explained by the effective theory at a smaller distance. In a sense, a given effective theory can be explained in terms of shorter-distance theories and the input from them. Dirac said that his equation that unified special relativity and quantum theory for the interactions of electrons and nuclei explained all of chemistry, and in a sense he was right. His equation, plus the input parameters describing electrons and nuclei, in principle explained all chemical processes. In another sense, however, he was not right, because in practice one could never start from the Dirac equation and calculate the properties of molecules, or figure out their structure, or determine how to construct new molecules with certain desired properties—the questions are just too complicated to solve. For example, Dirac could not have deduced that water is wet from his equation. For each effective theory, new regularities are found, and properties arise that are not predictable *in practice*. These are often called emergent properties. Life is an emergent property too. Physics tells us everything that molecules can do and cannot do. In particular, physics can tell us that life will not emerge on some planets if circumstances are too adverse, but it cannot guarantee that life *will* emerge on a planet where conditions are favorable, even though the emergence of life may be very likely.

Another way to view effective theories is in terms of types of understanding. At its own level, an effective theory provides a "how understanding," a description of how things work. But for the effective theory above it, at larger distances, the smaller-distance effective theory explains all or some of the input parameters, thus providing a "why understanding." For example, nuclear physics describes the properties of nuclei, using the proton and its electric charge, spin, mass, and magnetic properties as given, unexplained input. But the Standard Model provides the explanation, making it possible to calculate all those properties of the proton in terms of quarks bound by gluons.

Yet another perspective appears if we observe that in general, all areas of science are intrinsically open-ended: chemistry, the physics of materials, geology, biology, and so on. There is no end to the number of possible systems and variations that can be studied. But particle physics and cosmology are different. If the fundamental laws that govern the universe are found and understood, that's it—these two fields (that are merging into one) will end.

Having examined some implications of effective theories, let's return to the progression to smaller distances. We can go from atoms through the effective theory of nuclei, protons, and neutrons, to quarks and leptons. Another important point is that each effective theory works well at its level, but it breaks down as we go to smaller distances and find new kinds of structure. When we went inside protons we found quarks, so we could not make a theory of protons unless we understood quarks and their interactions as well.

To go deeper into matter, it will help to keep track of the distance scales numerically. Because we will cover a huge range of distances, we need to use powers of 10—remember that each step in the power is a factor of 10 in the result—10^{-1}\$ is a dime, 10^{-2}\$ is a penny, and 10^3\$ is one thousand dollars. There are two scales that are useful for us to keep track of: meters, which are a typical human size, and another length called the Planck length, after Max Planck, who first introduced it soon after he took the initial step toward the quantum theory in 1899. The Planck length is extremely small; we'll understand it better later in this chapter. People are typically a meter or two in size. All of our analysis will be very approximate, so we won't worry about whether we speak of the height or width or radius of a system. We'll keep track of powers of 10 but not worry about distinguishing between things that might differ by a factor of 2 or so in size. People are about 10^{35} Planck lengths (1 followed by 35 zeros) tall, 1 or 2 meters. Atoms are about 10^{-10} meter (one ten-billionth of a meter), about 10^{25} Planck lengths, in radius. Protons are about 100,000 times smaller than atoms, 10^{-15} meter or 10^{20} Planck lengths. Most particle physicists currently expect that quarks, leptons, photons, W and Z, and gluons will ultimately be understood as having a string-like extension if we could view them at a distance scale of about 1 Planck length, or 10^{-35} (a decimal point followed by 34 zeros and a 1) meter—they should seem point-like until we can study them at that scale.

In the language of this chapter, we can think of the Standard Model (described in Chapter 2) as the effective theory of quarks and leptons interacting on a scale of about 10^{-17} meter, or 10^{18} Planck lengths, about 100 times smaller than protons and neutrons. Sometimes we call this the collider scale, because it is associated with the typical energies at which the experimental collider facilities operate.

The goal of particle physics is an ultimate theory of the natural world. What should we call it? People have called it a *Theory of Everything*, but that name is somewhat misleading; it really is not a theory of weather, stars, psychology, and everything at once. It has been called the *final theory*; that is a good name, but when I first read it I misinterpreted it to mean the last in a succession of theories that replaced each other, as though all the theories on the way to the final one should be discarded. In fact, all the effective theories coexist simultaneously, and all are part of our description of nature. Therefore, I would like to choose a somewhat different name. I find I like calling it the *primary theory*, a term that suggests the theory one arrives at after going through a sequence of effective theories at smaller and smaller distances. As we will see more clearly in a few paragraphs, the primary theory should be the description of nature at a distance scale of about 1 Planck length, or about 10^{-35} meter. How can we journey the many orders of magnitude from the Standard Model to the Planck length?

Supersymmetry is an Effective Theory Too

If nature is indeed supersymmetric, one of the wonderful bonuses we may get is a way to carry out the journey through those orders of magnitude. Supersymmetry is an effective theory too, but it may be the penultimate one that will take us from the Standard Model to the "primary theory" near or at the Planck scale. Supersymmetry is an effective theory because it still needs some input parameters to describe the masses and interactions of the particles—those inputs should be predictable by the theory near the Planck scale, such as string theory. Qualitatively, the supersymmetric Standard Model should become the effective theory at distances of 10^{-17} to 10^{-18} meter, and it should remain the effective theory down to nearly the Planck scale. It has special properties that allow it to cover that large range, rather than breaking down at shorter distances as most effective theories do.

In the past we have been able to do the experiments that were essential to make progress as the technology developed and allowed us to probe more deeply. The Planck length is too small—direct experiments will never be possible at that scale. This statement is not just an extrapolation based on current technologies or costs. It is not just a matter of getting higher-energy probes. The probes have to have the energy concentrated into a region smaller than the scale of interest, and before we can do that, we run not only into limits like cost but also into natural limits. Nevertheless, there are a variety of ways to test ideas about Planck scale physics (this is explained further in Chapter 10). There are already some indirect ways to test ideas about physics at the Planck length, but supersymmetry will allow us to add many systematic tests. It will give us the techniques to take a prediction at the Planck length and calculate what is predicted at the distances colliders can reach in the coming years (about 10^{-18} meter) or to take data from colliders and calculate the form of the theory at the Planck length implied by the data. With supersymmetry we will be able to test ideas about string theories (Chapter 9) or whatever form the primary theory may take; without it we do not know how to do that. Of course, that bonus does not guarantee that nature is indeed supersymmetric, but it is a powerful motivation to study the theory and do the experiments needed to find out.

THE PHYSICS OF THE PLANCK SCALE

Whenever we describe a segment of nature, we have to talk about the actual quantities that are calculated or predicted or explained in units—meters, seconds, kilograms, or other appropriate units. In Carl Sagan's novel *Contact*, a signal from an extraterrestrial intelligence has been detected, with instructions on how to build a Machine to facilitate communication. There is a conversation between a scientist and an administrator.

> "Don't ask why we need two tons of erbium. Nobody has the faintest idea."
> "I wasn't going to ask that. I want to know how they told you how much a ton is."
> "They counted it out for us in Planck masses. A Planck mass is—"
> "Never mind, never mind. It's something that physicists all over the universe know about, right? And I've never heard of it."

For every effective theory there is a natural system of units, one where the description of phenomena is simple and not clumsy. It would be silly to measure room sizes in Planck lengths just because the primary theory is best talked about in those units. Consider the units for atoms more closely. The radius of an atom can be expressed in terms of the properties of the electron plus Planck's constant h, which sets the scale of all quanta. h is the fundamental, universal constant of quantum theory. Denoting the electric charge of the electron by e, and the mass of the electron by m_e, we find that the radius (R) of the hydrogen atom, the simplest atom, is equal to h^2/e^2m_e. The size of the atom is fully determined by these inputs. Nothing else matters. The nucleus, for example, is just a tiny object at the center. Once we know R, we can express the sizes of all atoms in terms of R; we don't need to use the input of h or the electron properties any more. R is the natural size unit for atoms. Atoms with different numbers of electrons will have somewhat different sizes, with radii such as $1.2R$ or $2.4R$, but all will be some number that is not too big or small times R. R is expressed in terms of parameters that are givens for atomic physics—maybe e and m_e can be calculated someday in string theory, but they cannot be understood in atomic physics.

We can learn a great deal from this kind of analysis. For example, this expression for the size of an atom has major implications. It tells us that the size of atoms is essentially a universal quantity. Given the basic quantities (Planck's constant and the mass and charge of the electron), the size of all atoms of all kinds, anywhere in the universe, is determined. Because mountains and plants and animals are all made of atoms, their sizes are approximately determined by the size of atoms and the electromagnetic and gravitational forces. Combining atoms into genes and cells to evolve an organism that can manipulate and deal with the world requires a large number of cells and sets a minimum size for the organism. Having a brain with enough neurons to make enough connections to make decisions about the world requires a brain of a certain minimum size, because the atoms cannot be made smaller than the size determined by the radius R. Nothing the size of a butterfly, anywhere in the universe, will ever be able to think.

Suppose now that we have just discovered the primary theory. To present the results, we have to express the predictions and explanations in appropriate units. What units should we use? We expect the natural units for the primary theory to be very universal ones, not dependent on whether the universe has people or stars. There is only one known way to make universal units. There are only three universal constants in nature common to all aspects of nature—to all interactions and all particles. They are Planck's constant h; the speed of light (denoted by c), which is constant under all conditions; and Newton's constant G, which measures the strength of the gravitational force. Because Einstein proved that energy and mass are convertible into one another, and gravitation is a force proportional to the amount of energy a system has, everything in the universe feels the gravitational force. In fact, using these three quantities—h, c, and G, it is possible to construct combinations that have the units of length, time, and energy. We expect all the quantities that enter into the primary theory or are solutions of the equations of the primary theory to be expressible in terms of the units constructed from h, c, and G. (For the interested reader, the result for the Planck length is $(hG/c^3)^{1/2}$, which is about equal to 10^{-35} meter, as we said before. For completeness, the Planck time is $(hG/c^5)^{1/2}$, which is about 10^{-44} second, and the Planck mass is $(hc/G)^{1/2}$, which is about 10^{-8} kilogram.) The Planck distance and time are extremely small, whereas the Planck mass (or, equivalently, energy) is very large for a particle.

Max Planck understood fully the universality of those units we call the Planck units. He wrote in his book *The Theory of Heat Radiation* (reprinted by Dover Publications, 1991),

All the systems of units which have hitherto been employed ... owe their origin to the coincidence of accidental circumstances, inasmuch as the choice of the units lying at the base of every system has been made, not according to general points of view which would necessarily retain their importance for all places and all times, but essentially with reference to the special needs of our terrestrial civilization. Thus the units of length and time were derived from the present dimensions and motion of our planet.... In contrast with this it might be of interest to note that ... we have the means of establishing units of length, mass, time ... which are independent of special bodies or substances, which necessarily retain their significance for all times and for all environments, terrestrial and human or otherwise, and which may, therefore, be described as "natural units." The means of determining the units of length, mass, and time ... are given by the constant h, together with the magnitude of the velocity of propagation of light in a vacuum, c, and that of the constant of gravitation, G. These quantities retain their natural significance as long as the law of gravitation and that of the propagation of light in a vacuum [and quantum theory] remain valid. They therefore must be found always the same, when measured by the most widely differing intelligences according to the most widely differing methods.

(I have left out some words to make this read smoothly, replacing them with ellipses, and because Planck didn't then know about the completion of the development of the quantum theory, I have added that term in brackets as he would presumably have included it.)

The Planck length and time can also be interpreted as the smallest length and time that we can make sense of in a world described by quantum theory and having a universal gravitational force. The arguments that teach us that are interesting and not too complicated, but to explain them we have to recall the definition of a black hole. Basically, the idea of a black hole is simple. Imagine being on a planet and launching a rocket. If you give the rocket enough speed, it can escape the gravitational attraction of the planet and travel into outer space. If you increase the mass of the planet, you have to increase the speed the rocket requires to escape. If you increase the mass so much that the required speed exceeds the speed of light, then the rocket can't escape because nothing can go faster than light. The rocket (and everything) is trapped. Light also feels gravitational forces, so beams of light are trapped too. Because gravitational forces increase with decreasing distance, if you pack some mass into a sphere of smaller radius, it is harder to escape from it, so the condition for having a black hole depends on both the amount of mass and the size of the sphere you pack the mass into.

Now the fascinating thing is that if we put an object having the Planck energy in a region with a radius of the Planck length, we satisfy the conditions to have a black hole! We cannot separate such a region into parts, or get information out from a measurement, so we cannot define space to a greater precision than the Planck length! Because distance is speed × time, and speed can be at most the speed of light, and there is a minimum distance we can define, there is also a minimum time we can define—that comes out to be the Planck time. We saw above that the Planck scale provides the natural units for expressing the primary theory when the units are constructed from the fundamental constants h, c, and G. Now we see a second reason for expecting the Planck scale to be the distance scale for the primary theory: There does not appear to be a way, even in principle, to make sense of smaller distances or times. The times when events occur cannot be specified, or even put in order, more precisely than the Planck time.

There is a third interesting argument that gives the same answer. The gravitational force between two objects is proportional to their energies and grows as the distance between them decreases. Consider, for example, two protons. Normally, the repulsive electrical force between them is much larger than the attractive gravitational force. But if the energies of the protons are increased to the Planck energy, then the gravitational force between them becomes about equal to the electrical force between them. All the forces become about the same strength at the Planck scale, rather than being widely different in strength as they are in our everyday world. Thus we might expect the gravitational force to unify with the others at the Planck scale, just as one might hope for in the primary theory.

The arguments of this chapter have led in several ways to the idea that it makes sense to analyze the physical world with effective theories organized by the distance scale to which they apply and to move toward a primary theory that unifies the forces and particles and is valid at the smallest scale that makes sense, the Planck scale. Of course these arguments do not prove that is how nature works; we will not know that until we achieve such a description. The Planck scale is very small, but not beyond our imagination. Some readers may recall the delightful *Powers of Ten* book and movie by the designers Charles and Ray Eames, in which the universe was looked at in snapshots each 10 times smaller than the previous one, starting with the largest cosmological distances. When the Eameses did this work, shortly before the Standard Model was discovered, they could not meaningfully go to smaller distances than the proton. Today the Standard Model takes us nearly 3 powers of 10 smaller than the proton. From the universe down to the Standard Model domain is about 46 powers of 10, and from the Standard Model to the Planck scale is only about 16 more powers of 10. Looked at that way, perhaps it does not seem so far.

Effective Theories Replace Renormalization

It is interesting to note that the approach of effective theories replaces and explains an old problem in particle physics. Readers who have followed particle physics in the past, or who have read treatments that are not up to date on this, may be interested in this brief section, though it is not needed for the rest of the book. When theories of particle interactions were studied in the 1930s to 1950s, it was found that some calculations apparently gave infinite answers. A procedure called renormalization was invented to solve the problem. This procedure was technically satisfactory and based on good physics, but it was conceptually weak. From the viewpoint of effective theory, the problem and the renormalization solution can be understood. Each effective theory breaks down as one takes it to shorter distances, or larger energies. For each effective theory, we have to input parameters from effective theories at shorter distances, such as the masses and charges. That input process is basically the renormalization procedure. Now we understand that as we go to smaller distances, or higher energies, we *expect* each effective theory to need renormalization—this is not a problem or an unexpected failure of the theory. However, the primary theory had better not need such inputs, or renormalization. It has to be a finite theory (one that never gives an infinite prediction for a physical quantity).

THE HUMAN SCALES

Since the time of Copernicus, who taught us that the earth is not at the center of the universe, we have learned that if we want to understand the world, we have to go beyond how the world seems to be and ask for evidence of how it is. We have learned that matter in the heavens and matter on earth obey the same natural laws, that we are made of the same atoms as the earth and the stars, that we and all organisms on earth evolved from cells, that we have unconscious minds that affect our behavior, and that our star is only one of a hundred billion stars in our galaxy. We have learned that the rules that govern nature (quantum theory and special relativity) are not apparent in our everyday classical world, and that the laws of nature have symmetries that are hidden from us but important (such as the particle interchange symmetry of the Standard Model). Perhaps even the number of space dimensions of the world will be larger than the three we are aware of (Appendix D). To understand the universe, we must recognize that additional hidden aspects of nature may arise at scales far different from the human scale, and we must learn how to uncover them. Supersymmetry may be such a hidden aspect of nature.

4

Supersymmetry and Sparticles

Finally, after three chapters of preparation, we can turn to examining the idea of supersymmetry. We have learned that the Standard Model's description of nature has two sorts of particles: (1) the basic constituents of matter, quarks and leptons, all fermions, and (2) the particles that mediate the forces, the photons and gluons and W and Z and graviton, all bosons. Fermions and bosons are treated very differently in condensed matter and atomic and nuclear physics, and in the Standard Model.

What Is Supersymmetry?

Supersymmetry is the idea, or hypothesis, that the equations of the primary theory will remain unchanged even if fermions are replaced by bosons, and vice versa, in those equations in an appropriate way. This should be so in spite of the apparent differences between how bosons and fermions are treated in the Standard Model and in quantum theory. The replacement occurs in the equations, of course, not for a fermion or boson in the real world.

Whenever an extension of the existing description of nature is proposed, it must pass many obstacles before it has a chance of being determined valid. It must be consistent with the rules of quantum theory and special relativity, and it must not change any of the tested consequences of the Standard Model. Because the fermionic or bosonic nature of particles derives from their spin, and spin is related to quantum theory and special relativity, both of which in turn involve space and time in their formulation, the formulation of supersymmetry must also involve space and time as well as the interchange of bosons and fermions. It is remarkable that such a property can be introduced without coming into conflict with some already established result.

The reader may be underwhelmed. Who cares whether the primary theory is invariant when its fermions and bosons are interchanged? OK, it's remarkable that this can be done in a relativistically invariant quantum theory, but so what? Actually, that's the way physicists felt about it too, for the most part, originally. The supersymmetry theory was technically very beautiful, so theorists enthusiastically studied it, but it was spoken of as a "solution in search of a problem." I myself got into the field because I asked at a number of talks in the late 1970s how we would know if nature actually was supersymmetric, and the typical answer was that no one had thought very much about that. It was such a beautiful theory that I began to feel strongly that it was important to test it experimentally. I still feel that way.

© 1994 by Sidney Harris

My colleague Michael Duff recalls two incidents that occurred while he was a young Lecturer at Imperial College, London. In 1979 the theory group there applied for funding to support research activities, particularly postdocs. Their request was approved, contingent on the funds *not* being spent on supersymmetry research. A couple of years later, Duff applied for support to attend a meeting on supergravity that Stephen Hawking was organizing in Cambridge. The request was rejected, with an explanation that such research was not deemed a suitable use for funds in particle theory.

SOME MYSTERIES
SUPERSYMMETRY WOULD SOLVE

There were indeed reasons to be excited about supersymmetry, but it wasn't obvious at the beginning. That's related to how supersymmetry arose, as an idea, in a manner different from other ideas in the history of science. New ideas, including the Standard Model, had always come as a response to trying to understand observed regularities, or to puzzles, or to a desire to explain the behavior of aspects of the natural world, or to apparent inconsistencies in existing descriptions. Supersymmetry, by contrast, was originally noticed as a property of certain models (with fewer than three space dimensions) being studied for their own sake. A fascinating aspect of its history is that supersymmetry was not introduced to solve any experimental mystery or resolve any theoretical inconsistency. As it was studied and came to be better understood over a decade, theorists realized that it actually could solve a number of the important mysteries in particle physics and could provide new approaches to others. For many physicists, that supersymmetry solved problems it was not introduced to solve was itself a powerful hint that it was indeed part of the description of nature. No other major concept in physics has ever been introduced as an idea not related to any data or inconsistency and then been found, later, to solve important problems.

In later chapters we will examine some of these mysteries and their solutions in a supersymmetric world. Here I will list them with a brief explanation to give an overview of the impressive impact of the supersymmetric theory. I have been using the word *mystery* in the detective story sense. We expect mysteries to have solutions and to be solved—usually by the end of the book. The following list should indicate why many physicists work on the theory of supersymmetry or on testing it experimentally and why nearly 10,000 papers on supersymmetry have been published *prior to* the discovery of any explicit evidence that it is really how nature works.

• In order to incorporate the masses of the particles into the Standard Model description, physicists postulated the existence of a Higgs field. They also assumed that this field interacted in a very specific and somewhat enigmatic way. The Higgs physics was a mysterious part of the Standard Model for most physicists, hard to accept and hard to test, though technically it solved the problem it was introduced for. In particular, some new physics beyond the Standard Model *must* exist to provide the Higgs physics—the Standard Model simply cannot lead to an explanation of the Higgs interaction needed to account for the masses of particles in a consistent way. Then in 1982, several theorists figured out that if the Standard Model were extended to be supersymmetric it could provide an elegant physical explanation for the Higgs physics. For many theoreticians this was the result that convinced us that supersymmetry was a property of nature, not just nice mathematics. Even better, in order for the supersymmetric approach to Higgs physics to work, it was necessary that the top quark (whose mass was not measured until the 1990s) be unusually heavy compared to the other quarks and leptons. That the top was predicted to be heavy, and was indeed confirmed to be heavy a decade later by data, was a powerful indirect test of the validity of supersymmetry.

• The Standard Model has a very serious conceptual problem called the hierarchy problem. We saw in Chapter 3 that the natural scale for the primary theory was the Planck scale, about 10^{-35} meter. The Standard Model is a description of quarks and leptons and their interactions, at a scale of about 10^{-17} meter. The problem is that in a quantum theory, physics at every scale may contribute to physics at every other scale, so it may not be consistent to have these two scales so separate; rather, the Standard Model scale and the Planck scale should be very near each other, and in fact the Standard Model scale should be near the Planck scale. Another way to view this problem comes from recognizing that in the Standard Model, all the masses of electrons, quarks, W's, and Z's should be either zero or the Planck mass. For the Standard Model this is indeed a major problem, even though it is a conceptual problem that does not explicitly affect the experimental predictions of the Standard Model. The problem has two parts. First, given that there is a separation of the Standard Model scale from the Planck scale, why does the Standard Model end up where it is (at about 10^{-17} meter) and not at some other scale? Second, and more important conceptually, what can make the theory maintain that separation in a mathematically consistent way? The supersymmetric Standard Model solves the second problem and gives insight into the first. It does so in a manner that uses the unification of fermions and bosons in an essential way. The very nature of fermions and bosons implies that they contribute to the coming together of scales in ways that cancel, so the mixing of scales can be canceled in a general way, which solves this problem.

• For two centuries, physicists have been actively trying to unify our description of the forces of nature. Having five different forces, rather than one basic force, suggested we were missing some unifying principles. Maxwell succeeded in unifying electricity and magnetism, and the Standard Model unifies the description of weak interactions and electromagnetism, so there has been some progress. In a quantum theory, one can calculate how a force would behave if one could study it at smaller distances. Remarkably, when this is done in the Standard Model for the electromagnetic, weak, and strong forces, they become more and more like each other at shorter distances, though finally they do not quite become equal in strength at any distance. More remarkably, when the study is repeated with the supersymmetric Standard Model, as was done in the early 1980s, forces *do* become essentially equal at a very small distance, about 100 times larger than the Planck scale. That did not have to happen—nothing in the Standard Model implies that the forces should become equal. And if they do become equal, nothing in the Standard Model implies that this should happen at just the distance scale that would be expected if full unification occurred at the Planck scale. Presumably these are important clues toward a better understanding. Because we do not know the form of the complete theory at such small distances very well, we do not know how to extrapolate to even smaller distances, but some reasonable ways of doing so suggest that these forces become equal to the gravitational force at about the Planck scale, a result that much encourages the idea that the goal of understanding the forces of nature in a simple way will indeed be reached. The superpartners seem to be needed for this to work.

- All of the superpartners are expected to be unstable particles, decaying into lighter superpartners, except for the lightest superpartner (LSP), which has no lighter ones to decay into and is accordingly stable. Thus supersymmetry introduces a new stable particle into the universe, joining photons, electrons, neutrinos, and protons. The light we see from stars is composed of photons. Protons and electrons form the stars and planets. Neutrinos, and the LSP (if it exists), will be forms of matter that are present throughout the universe. Because they feel only the weak and gravitational forces, not the electromagnetic or strong forces, they will not participate in forming stars. They will be "dark matter." Supersymmetry predicts that there is dark matter composed of the LSP. Right after the Big Bang, there were about the same number of each kind of particle. Most particles decayed into lighter ones, and some annihilated into others as the universe expanded and cooled. We have a theory of how they all interact, so we can calculate how many are left now. Even though they did not coalesce into stars that produce photons that we can see, their presence can be detected by their gravitational attraction for what we do see, if there are enough of them—their presence modifies how stars move in galaxies and how galaxies move relative to one another. It was realized in the early 1980s that supersymmetry predicted that there should be considerably more LSP dark matter than even the matter in stars. Indeed, astronomers had already observed that the universe did have considerable dark matter, because stars and galaxies did not move through the universe as they would if the only matter was what we could see, but at that time it was not known experimentally whether the dark matter could be nonluminous forms of ordinary matter (such as interstellar dust) or whether a previously unknown kind of matter must exist. (We will examine the dark matter and how to study it in the laboratory in Chapter 6.)

- The LEP collider at CERN and the SLC collider at Stanford were constructed during the 1980s in order to test the Standard Model and to search for new physics that would strengthen the foundations of the Standard Model. Before LEP and SLC operated, it was possible to predict the kinds of results they should find in a supersymmetric world. Either superpartners would actually be produced and detected, or the effects of supersymmetry on the quantities measured would be so small that observables should have essentially the values the Standard Model predicts for them. Nonsupersymmetric approaches to an explanation of the Higgs physics generally predicted larger effects. After a decade of accurate measurements, no superpartners were produced, unfortunately, but the results did confirm the prediction that there were no sizable deviations from the Standard Model expectations. (If you have sufficient confidence in indirect arguments, you might even conclude that these two considerations—the need for some extension of the Standard Model to explain the Higgs physics, and the absence of any deviations from the Standard Model expectations for observables in the LEP and SLC data—together confirm that supersymmetry must be a part of the description of the world. But most physicists want verification that is less indirect.)

- We have already seen that supersymmetry suggests, in two (consistent) ways, that the description of the electromagnetic, weak, and strong forces will be unified with gravity. First, all these forces became of the same strength at very small distances near the Planck scale. Second, supersymmetry, as a symmetry throughout space and time, necessarily had a connection to the theoretical description of gravity (more on this below). Although the connection between supersymmetry and gravity is not yet fully understood, it is very encouraging that the supersymmetric Standard Model is related to the theory of gravity, whereas the Standard Model itself is not.

- In Chapter 9 we will examine the connections between string theory and supersymmetry. String theory seems to require supersymmetry as an essential part of the theory, though we do not yet know whether this is truly a necessary condition. In order to learn how string theories work, and whether their solutions might really describe our world, we must learn how to solve the equations of the theory. Finding solutions has turned out to be much easier (and therefore sometimes possible) because of the powerful constraints that have to be satisfied when the theory is supersymmetric.

• Supersymmetry has led to new approaches to solving or explaining a number of other important problems that had no possible solution in the Standard Model. Here I will just list some, without explaining them, to give a sense of the opportunities. In Chapter 8 we will examine several of them. Fundamental issues for which supersymmetry provides new ideas and methods of approach include how the universe got to be mainly matter and not antimatter, whether and how protons decay, why the universe is the age and size it is, and the rare decay of quarks and leptons.

• Finally, probably the most important consequence if supersymmetry is part of our description of nature is that it provides a window through which to look at the Planck scale from our world that is so distant from the Planck scale. We have seen that there are several reasons to expect that the primary theory will be naturally formulated at the Planck scale. But we cannot hope ever to do actual experiments at the Planck scale—it is just too small. If supersymmetry is part of our description of nature, it implies that we can write the predictions of a candidate for the primary theory at the Planck scale and then calculate what should be observed in experiments we can do at colliders, for neutrino masses, for proton decay, and much more. Similarly, we can measure parameters that are part of the description of the theory in experiments and then calculate the values they have at the Planck scale (many of them have values that depend on the distance scale). It is most likely that the latter procedure will be the actual one, with experimental input needed before it is possible to formulate detailed candidates for the primary theory. We can hope someone will be smart enough to guess the primary theory, but if not, we can get there anyhow by building up knowledge about the form the theory must take, using experiment and theory together as physicists have traditionally done so well. Even if someone does guess the primary theory, they won't be sure—and no one will believe it—unless there is experimental input of the kind supersymmetry can provide. If the world is not supersymmetric, we do not know of any way to relate physics at the Planck scale to physics at our scale. If the world is supersymmetric, we can expect both to be able to formulate the primary theory and to test it.

The above arguments were of two kinds. Some provided explanations for actual phenomena, or predicted observations, whereas others were about our opportunities to make sense of the world at the deepest levels. The first kind provide real evidence that the world is indeed supersymmetric— indirect evidence, but still real and significant evidence. The second kind, of course, does not imply that supersymmetry is real—just because it will help us solve string theories or provide a window on the Planck scale does not mean nature *is* that way. But both kinds of reasons contribute to the very strong and widespread interest in supersymmetry among physicists. And it is worth repeating that supersymmetry was not invented for any of these reasons; all the reasons emerged as its implications were studied. Indeed, the results described above provide impressive examples of emergent properties of a basic theory, properties implied in a sense by the theory but not apparent until suggested by data or after much study.

Supersymmetry has the consequences it does because the requirement that the theory be unchanged when bosons and fermions are interchanged can be satisfied only if the theory (the equations) is constrained to have a certain form. It may be helpful to mention a couple of historical examples of properties that emerged as theories were more constrained. Maxwell took several equations that people had developed to explain electrical and magnetic phenomena and tried to combine them into one consistent set. He found that to do so, he had to add a term to one equation. The new set of equations then turned out to have a new solution that described light as an electromagnetic wave—suddenly the behavior of light was incorporated into the electromagnetic theory, and the electromagnetic waves we now use for communication were unexpectedly predicted to exist. Another example comes from special relativity. Einstein found that to have the same laws of electromagnetism hold in all systems shifted in position or speed relative to one another, it was unavoidable that space and time be tied together. In any mathematical science, new constraints on the basic equations often imply major new physical phenomena, just as we have seen happened in the case of supersymmetry.

The Superpartners

In Chapter 2 we saw that an important aspect of the Standard Model was the invariance of the theory under the interchange of certain particles, the electron and the electron neutrino, the up and down quarks, and so on. Some of the particles required for that invariance to be valid were already known to exist when the Standard Model was developed, but others were later predicted to exist and then, later still, were found. We also saw that having the theory be consistent both with quantum theory and with special relativity required the existence of antiparticles. None of the antiparticles had been observed at the time they were predicted, but because physicists have learned what to look for, all have been observed since then. For supersymmetry to be valid, it would again be necessary for previously unknown particles to exist that are like the particles of the Standard Model. Because bosons and fermions have different spins, the partners must differ in their spins. There must be another particle just like a photon, with no electric charge or weak charge or strong charge, but with spin one-half instead of spin one. There must be another particle just like the electron, with the same electric charge and weak charge as the electron and with no strong charge, but with spin zero instead of spin one-half. An examination of the particles we know shows that none of them have the properties to be the needed partners. The situation is analogous to what happened with antiparticles, where all the predicted new particles had yet to be found. We call the new set of particles *superpartners* or *sparticles*. The sparticles form smatter.

As our understanding of nature progressed through the past century, we saw the number of "electrons" grow, in a sense. From 1895 until the 1920s, there was thought to be only a single electron. Then first the data about energy levels of atoms, and later Dirac's unification of quantum theory and special relativity, led to assigning half a unit of spin to the electron. In the quantum theory, that meant the electron spin could point up or down with respect to any arbitrary direction and accounted for the observed extra energy levels of atoms. It is as though we should think of two electrons, one with spin up and one with spin down. Then antiparticles were predicted and found. Again, it is as though electrons had to come in four kinds: particle or antiparticle each with spin up or spin down. With the validation of the Standard Model in the 1970s, we learned that electrons and neutrinos were really different projections of the same basic particle that could turn into each other by emission or absorption of a W boson, and that there were analogous processes for antielectrons and antineutrinos, so in a sense we think of one object that can exist with spin up or down and as electron or antielectron, neutrino or antineutrino. With supersymmetry we go one step further. Each of these states of the "electron" is predicted to have a partner that differs only in its spin being zero rather than one-half.

There might seem to be a proliferation of particles. But if the theory says that given any one of them, all the others must exist, then the apparent proliferation follows from a conceptually simple structure. Just as a simple structure, based on the electron and on the up and down quarks and their interactions, underlies the great complexity of our world, so there is a simple conceptual structure with a small set of Standard Model particles plus their antiparticles and superpartners.

There is a simple notation and terminology for superpartners. The superpartner of every particle is written as the partner with a tilde ($\sim$) over it: $e \to \tilde{e}$, γ (photon) $\to \tilde{\gamma}$, etc. If the Standard Model particle is a fermion, then the name of the partner is the fermion name with an s- in front: sfermion, selectron, squark, up squark, stop, sneutrino, etc. If the Standard Model particle is a boson, then the name of the partner is the boson name with the suffix –ino added or replacing the –on: photino, gravitino, Wino, higgsino, etc. Because all of the regularities of the Standard Model hold for the supersymmetric Standard Model too, keeping the names helps in keeping track of processes and predictions. Table 4.1 shows the particles and their superpartners. As many of us have said, we have a new slanguage spoken by sphysicists.

It should be emphasized that supersymmetry is the *idea* that the laws of nature are unchanged if fermions $\leftrightarrow$ bosons. It is not that an electron becomes a selectron in the equations that constitute the basic theory but, rather, that the equations should contain symbols representing both electrons and selectrons, and the equations are unchanged if those symbols are interchanged. The existence of the sparticles themselves is the most dramatic prediction to test that idea—they are the smoking gun. There are several other tests that are less explicit. Some indirect tests based on the implications of supersymmetry have already been positive for supersymmetry, helping

TABLE 4.1—Particles and Superpartners

Particle	Name	Feels These Forces[a]	Mediates These Forces[b]	Superpartner
e, μ, τ	charged leptons (electron, muon, tau)	EM, W	—	sleptons $\tilde{e}, \tilde{\mu}, \tilde{\tau}$ (selectron, smuon, stau)
ν_e, ν_μ, ν_τ	neutrinos	W	—	sneutrinos $\tilde{\nu}_e, \tilde{\nu}_\mu, \tilde{\nu}_\tau$
u, c, t	up, charm, top quarks	EM, W, S	—	squarks $\tilde{u}, \tilde{c}, \tilde{t}$
d, s, b	down, strange, bottom quarks	EM, W, S	—	squarks $\tilde{d}, \tilde{s}, \tilde{b}$
γ	photon	[c]	EM	photino[d] $\tilde{\gamma}$
$W^\pm$	weak boson	EM, W	W	Wino[d] $\tilde{W}^\pm$
Z	weak boson	W	W	Zino[d] $\tilde{Z}$
g	gluon	S	S	gluino $\tilde{g}$
G	graviton	GR	GR	gravitino $\tilde{G}$
h	Higgs boson[e]	W	generates mass	higgsino[e] $\tilde{h}$

[a] All particles feel the gravitational force.

[b] EM = electromagnetic force, W = weak force, S = strong force, GR = gravitational force.

[c] Photons feel only the gravitational force, but they interact with all electrically charged particles.

[d] Mixtures of these particles form charginos and neutralinos (Appendix C).

[e] The additional Higgs bosons predicted by supersymmetry are not shown.

convince many physicists that it is indeed correct. This indirect evidence is outlined earlier in the chapter, and we will examine some of these tests in detail in later chapters.

Supersymmetry requires not only new (to us) particles but also new interactions involving the new particles. It is easy to write the new interactions once one has the Feynman vertices of the Standard Model as shown in Figures 2.1 and 2.2. To make the theory supersymmetric, simply take each of the vertices and add to it those obtained by replacing pairs of particles by their superpartners. (That an even number of particles has to be replaced by superpartners follows from the properties of spin.)

For example, the vertices of Figure 4.1 (and analogous vertices for other particles and their superpartners) are added to that of Figure 2.1 in the supersymmetric Standard Model. I didn't draw arrows on these lines for simplicity; the arrows can be put on the lines in any directions as long as the total amount of electric charge does not change in the interaction. Just as before, more complicated diagrams are made by joining vertices in all possible ways, and the rules tell us how to compute the probability of any resulting process.

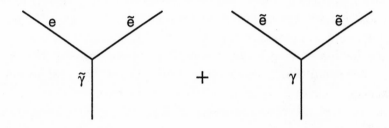

FIGURE 4.1.

The theory implies that the interactions of Figure 2.1 and of Figure 4.1 are of the same strength; this important prediction will be a key test of the theory if and when the selectrons and photinos and other sparticles are observed. Analogous replacements are made for the other vertices of Chapter 2. All processes that can happen in a world described by the supersymmetric Standard Model are then obtained by joining the vertices in all possible ways. The rules of quantum theory tell us how to calculate the expected rates of each sprocess from the sfeynman diagrams.

SUPERSYMMETRY AS A
SPACETIME SYMMETRY: SUPERSPACE

We have seen that there are two important ways to think of supersymmetry. They are consistent, of course, but often one or the other is more helpful in our efforts to achieve a given goal. The first and most common way to think of supersymmetry is as a symmetry of the laws of nature under the interchange of bosons and fermions. A sparticle is predicted to exist for each of the quarks and leptons and bosons of the Standard Model. The experimental approach to supersymmetry and most practical calculations are carried out in this framework. A second approach is thinking of supersymmetry as an effective theory that opens a window on the Planck scale, so that we can formulate and test the primary theory; this approach shows us how important supersymmetry is in our search to understand our universe. And there is a third way. Sometimes it is fruitful to think of supersymmetry as a spacetime symmetry, but in an extended spacetime called superspace. This approach is often most helpful in uncovering the mathematical properties of the supersymmetry theory. The remainder of this section is a little more technical than the rest of the book, and you do not need it to follow the rest.

Think about an isolated particle or object, one on which no forces act. Assume that it is moving, with some energy and momentum. If there were forces, they could change its energy and momentum, but with no forces we expect the energy and momentum to be unchanged. We say the energy and momentum are conserved. We are making assumptions when we assume that energy and momentum are conserved in the absence of forces. We expect space and time to be neutral, so the object can move through them without feeling effects. If no other objects are around, we don't think it matters whether our object is here or a few miles away, or whether it is rotated some amount, because we think space and time don't affect objects in them. Whenever something is invariant under changes, we speak of a symmetry. Whenever some quantity is conserved, there is an associated symmetry, and vice versa. Just as we can change our position in space or time and find symmetries, we can change our position in superspace and find an associated symmetry—supersymmetry.

Symmetries in physics can be of two kinds. The examples we just discussed are "geometrical" or spacetime symmetries. In Chapter 2 we saw that the Standard Model has some "internal" symmetries, ones for which the theory is invariant under interchange of particles such as up and down quarks. A very important question is whether there could ever be a symmetry that mixed geometrical and internal symmetries. For example, we can be pretty sure that moving an up quark a few centimeters won't change it into a down quark, but maybe there is some more subtle geometrical change that could. It turns out that for our standard ideas of spacetime, the naïve approach is right; geometrical and internal symmetries cannot be mixed. It can be proved that the spacetime symmetries we already know about are the only possible ones. However, if we allow for new "fermionic" dimensions, then it turns out that one more symmetry can exist, and it is supersymmetry. Another attractive feature of supersymmetry is that it is the last possible spacetime symmetry nature could have, the only mathematically possible symmetry that is not already known to be part of nature.

Supersymmetry is the idea that the basic laws are invariant under interchanging fermions and bosons. We just saw that invariances lead to symmetries. In our usual spacetime, bosons and fermions behave differently. It turns out that we can attach to our normal spacetime another four "dimensions" that will allow us to incorporate differences between fermions and bosons into the structure of the "space," defining a "superspace." Superspace is a geometrical structure in which fermions and bosons can be treated fully symmetrically.

The extra superspace dimensions are not like our dimensions. They have no size at all. Nor are they like the extra small dimensions of string theory (Appendix D). In the superspace, we can think of bosons and fermions as two different projections of a single object, much as an electron and its neutrino can be thought of as two different projections of a single object in an internal space. Superspace is an intrinsically quantum theoretical structure, because the whole idea of fermions is meaningful only in a world described by quantum theory. One can think of superspace as adding fermionic dimensions to our usual bosonic dimensions. If the world really exists in superspace, it affects observables—it is a testable idea. The simplest and most dramatic test is that the superpartners must exist. If superpartners are found, we can interpret the results as letting us learn experimentally about the properties of superspace.

One can turn the history around. Suppose that quantum theory had been formulated from the beginning in superspace. Then immediately it would have predicted that all particles would exist in both a bosonic and a fermionic form. Supersymmetry, formulated in superspace, requires that fermions exist—in a sense, it explains their existence.

Einstein's general relativity is a geometrical theory of gravity, which views the gravitational force as an effect of the distortion of spacetime by masses (such as planets). Once superspace was formulated, people immediately thought of using it as the basis of a generalized geometrical theory of gravity, "supergravity." Supergravity incorporates general relativity and extends it. The graviton that mediates gravity is predicted to have a superpartner, the gravitino. The connecting of supersymmetry to gravity encourages people to think that the unification of gravity with the Standard Model forces will include supersymmetry. Interestingly, the geometrical structure of supergravity takes a simple and elegant form if the spacetime is extended to eleven spacetime dimensions (with associated superspace coordinates). As it happens, that is also the largest spacetime dimension in which one can write a consistent theory of gravity.

Hidden or "Broken" Supersymmetry

So far, I have argued that there is good reason to expect that the laws of nature are supersymmetric. In fact, however, the situation is more subtle. We know that nature is not exactly supersymmetric for two kinds of reasons. First, if the world had a selectron with every property identical to that of the electron except its spin, we would already have observed it in experiments. That's true for a number of other sparticles too. Second, even if the relevant experiments were not possible, we could deduce that the world would be entirely different if selectrons with the same mass as the electron existed. In fact, there would probably be no life in the world. That is because if bosonic electrons existed, they would all fall into the lowest energy level of an atom, shielding the electric charge of the nucleus, and there would be no valence electrons to bind atoms into molecules. (This reason should not be taken as an excuse for supersymmetry to be broken, but it illustrates how indirect analyses can be helpful in understanding how the world works.)

In physics, a system's symmetries can be powerful guides to how the system behaves. Some symmetries are exact ones; lots are not. But even when they are not exact, they still can be very helpful in our efforts to understand the behavior of the system. Take a very simple example—think of a child's spinning top. It has a cylindrical symmetry about the axis it spins around. Suppose it were poorly made, somewhat bumpy. It would still spin. Maybe it would wobble a bit if it were a little heavier on one side. It might spin a little less long and be more likely to end up lying on one side than another, but it is still recognizably and basically a top even though its cylindrical symmetry is not quite right. When a symmetry is only partly valid or somewhat imperfect, physicists speak of it as a broken or hidden symmetry. An example closer to our interests is the symmetry between particle and antiparticle found by Dirac from which he deduced the existence of antiparticles. That symmetry was called "charge conjugation invariance" because Dirac's equation did not change when the sign of the charge was reversed. It is a valid symmetry for the particles alone, and it is also valid for the ways the particles interact via the electromagnetic and strong forces, but it turns out it is not valid for interactions via the weak force. The important point is that despite the partially broken symmetry, all the predicted particles (the antiparticles) have been found to exist. The supersymmetry case is expected to be somewhat similar to the antiparticle case. All of the superpartners should exist, but they can have masses different from their particle partners. The interactions will be similar, but they can differ because the masses are different, and for other associated reasons. If the superpartners are heavier, they might be too heavy to have been observed in experiments so far.

The issue of hidden supersymmetry illustrates well how physics progresses. In the 1960s there was no theory of the forces or basic particles at all. Then the Standard Model emerged. It provided a very good description of the particles and interactions. The remaining problem was to determine what form the Higgs physics takes so that the particle masses could be incorporated into the theory and to understand from what underlying physics the Higgs physics itself arose. Supersymmetry in turn can provide the answer to the Standard Model Higgs physics problem: It explains how the Higgs physics arose. The supersymmetry answer actually depends on the way supersymmetry is hidden or broken, and it ties the supersymmetry breaking to the Higgs physics of the Standard Model (more about that in the next paragraph). We do not yet understand how the supersymmetry is broken. That should be explained by the next level of effective theory; probably the supersymmetry breaking won't be explained within the supersymmetric Standard Model itself. But supersymmetry is a sufficiently comprehensive theory that it is possible to know the form the broken-supersymmetry theory must take, even though we do not yet understand the mechanism that generates that form. (This is very much like the way we could incorporate the masses using the Higgs physics even though we did not yet understand how the Higgs physics arose.) A short-sighted view might claim we had just traded one problem for another, but that is really not so. First, the new, extended supersymmetric Standard Model incorporates a variety of phenomena that were mysterious before it emerged, and second, the supersymmetric Standard Model is one effective theory closer to the primary theory. If nature is indeed supersymmetric, explaining the origin of supersymmetry breaking will become the central problem of supersymmetry theory and experiment.

The manner in which supersymmetry explains the Higgs physics is elegant and has important consequences for how we expect to test supersymmetry experimentally. It is rather technical. A more detailed description is given in Appendix B; here I will give a short version. There are three parts to the Higgs physics of the Standard Model. First, the Higgs field must exist. Then it must interact in a certain way with the other particles—that is called the Higgs mechanism. Third, the quanta of the Higgs field, one or more Higgs bosons, must exist—that is required once the Higgs field exists. The supersymmetric Standard Model automatically contains fields like Higgs fields, so it makes their existence more natural. Supersymmetry also provides the interaction, the Higgs mechanism. The crucial signal that the Higgs mechanism is working is that a parameter of the theory that naïvely should be interpretable as the square of a mass (let's call it M^2 to give it a name), and therefore should be a positive number, is in fact negative. Although that seems at first to be a difficulty, it turns out on deeper examination to do what is needed to give the other particles a mass. Thus one can tell whether the Higgs mechanism is working in any theory by checking to see whether the quantity called M^2 is negative. The Higgs mechanism is described in more detail in Appendix A. In the supersymmetric Standard Model, M^2 is one of the supersymmetry parameters and has a positive value at the Planck scale, so all the quarks and leptons and W and Z are massless at that scale. But quantities such as M^2 aren't constant—rather, they vary as the theory is applied at different distance scales. We need to know M^2 at the scale of the weak interactions and larger scales, where we live and where our experiments show that the quarks and leptons and W and Z have mass. In the supersymmetric Standard Model, we can calculate how the value of M^2 changes as it goes from the Planck scale to the weak scale. The remarkable result is that if one condition holds, M^2 decreases to zero and becomes negative at larger distances, inducing the Higgs mechanism!

That condition was that one of the quarks had to be rather heavy, heavier than a W boson. When the way supersymmetry could explain the Higgs mechanism was first pointed out in the early 1980s, the heaviest known quark was far lighter than the W boson, but the mass of the top quark was not known. Therefore, supersymmetry predicted that the top quark would be far heavier than the naïve estimates. When the top mass was finally measured in the 1990s, it was indeed about twice the W mass, comfortably satisfying the supersymmetry prediction. If history had been a little different and supersymmetry had been developed and understood before the Standard Model, the Higgs mechanism would not seem mysterious at all but would simply be a new and unexpected consequence of supersymmetry.

The way supersymmetry explains the Higgs mechanism has an important implication: It is the only way we know to relate the masses of the superpartners to known masses so that we can estimate how large the superpartner masses are. The Higgs mechanism leads to a description of the masses of the Standard Model particles in terms of the assumed masses of the superpartners. Thus we obtain an equation with a known Standard Model particle mass (W or Z mass) on one side and unknown superpartner masses on the other. For any equation like that, we would not trust the result if the quantities on one side were much larger than those on the other, because any measurable quantity in a physics theory can only be estimated to some accuracy; there are always experimental and approximation errors involved. Therefore, the supersymmetric Standard Model explanation of the Higgs mechanism would not make sense unless the superpartner masses were not much larger than the Standard Model masses they explain. That gives us an estimate of the masses we should expect the superpartners to have as we search for them, and it tells us at what stage we should question the validity of the theory if the superpartners have not been detected. Such estimates are only approximate, but luckily the expected masses are small enough that they imply the superpartners should be detected soon. The next chapter describes this experimental search.

Testing Supersymmetry Experimentally

The main goals of collider experiments now are finding superpartners and Higgs bosons, or (if somehow we are on the wrong track in spite of the indirect evidence and strong theoretical arguments) showing that the superpartners and Higgs bosons do not exist. The Higgs bosons will both complete the Standard Model and provide clues about how to extend the Standard Model. The search for them is the focus of Chapter 7. This chapter explains how to recognize a superpartner and describes the facilities where they might be observed. Chapter 6 looks at the special properties of the lightest superpartner that make it a very good candidate for the cold, dark matter of the universe and discusses special ways to explicitly detect it.

DETECTORS AND COLLIDERS

How can we get experimental evidence about aspects of the world that are too small or too far away for us to see directly? That question would be a good unifying theme for a history of scientific discovery. We can start by asking how we see anything in our world. Photons bounce off an object and then enter our eyes, forming an image after significant processing by our brains. People first looked deeper than we can see with the unaided eye by making glass lenses that magnified images. Optical microscopes enable us to see things thousands of times smaller than the unaided eye can see, but not smaller than that—not even as small as an atom. When experiments reached that level, it became clear that understanding would require going to smaller distances. More energetic particles can probe more deeply. In the natural world, the most energetic particles we can control to study something are emitted from the decay of nuclei. They have an energy about a million times greater than that of visible photons, so essentially they permit us to see something a million times smaller. To probe even more deeply, we needed a way to increase the energy of particles; accelerators were invented in the 1930s, and they have been improving ever since.

Of course, we do not see the particles the same way we see photons. It is necessary to add an intermediate stage, a detector that collects the information about the behavior of the particles and then presents an image for us. Often it is an image we see, though sometimes even apparent images have more stages between the object and ourselves—for example, today telescopes often really show us digitized interpretations of light outside the visible range. The first experiments using beams of particles from nuclear decay revealed how they bounced off atoms by surrounding the target with screens that would fluoresce when hit by a particle. Then people observed the light from the fluorescing screens. The people were one step removed in the observation process. The information could have been recorded on film and studied later, which would have added more stages. Today, several stages are often present before the information is studied by a person. A camera is a detector—it sees by recording the effects of photons. The detectors that are used in particle physics have to also "see" electrons, muons, and other particles to reconstruct the information from the collisions, so they are much larger and more complicated than cameras.

In a sense, the detectors can be thought of as the world's biggest video games. A collision occurs. As much information as possible is extracted from the emerging particles. From the information they are able to extract, the players have to figure out whether the event is one expected in the Standard Model or is "new physics." One way players can win is to find new physics events. At Fermilab there are two collider detectors (one is shown in photo 5.1 on the next page). The probability of a new physics event occurring is the same for each of them; they compete to see which can find and confirm the new physics first. Actually, this is an imperfect analogy because a deeply serious and heroic effort is needed to design, construct, and use a detector to probe nature to a millionth of a billionth (10^{-15}) of what we can see—it is not

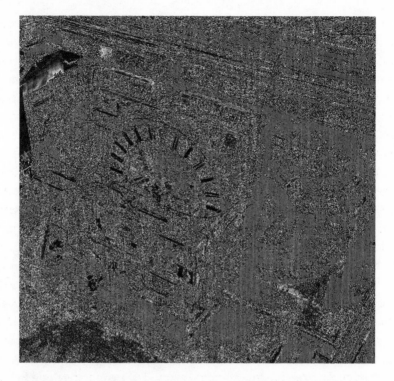

PHOTO 5.1.   A Fermilab detector (Fermilab photo).

a frivolous activity. The picture of a Fermilab detector above conveys a sense of the size and complexity needed to "see" nature at such tiny distances. The detector is the smallest "microscope" that would allow us to observe the superpartners. Perhaps it is a "superscope!"

Einstein taught us that energy and mass are in principle interchangeable, and eventually people realized that this provided a new way to get information about particles. If particles could be made very energetic and then made to collide with other particles, some of their energy could be converted into the creation of previously unknown particles. (Sometimes we'll call them new particles, even though they are new only to us, not to nature.) The first new particles were detected in the 1930s, when people set up detectors to observe cosmic rays that hit a nucleus in or near the detector and converted most of their energy into making particles. (Cosmic rays are particles impinging on the earth from outer space. Sometimes they are very energetic, most of the energetic ones originating from exploding supernovas.) Later, physicists learned to accelerate electrons and protons to higher energies with electric fields and eventually to cause them to collide so that even more energy was available—when a swung bat and a ball collide, the ball goes a lot farther (gets a lot more energy) than it would have if it had hit a stationary bat. In practice, as we will see in more detail below, only processes allowed by the Feynman diagrams and the rules of quantum theory can generate new particles, so the experiments have to be planned carefully. When particles are produced in a collision, they are not particles that were somehow inside the colliding particles. They are really *produced* by converting the collision energy into mass, the mass of other particles. Which particles will be produced is partly determined by their mass—the lighter they are, the easier it is to produce them, other things being equal—and also by the probabilities calculated from the Feynman diagrams. In this chapter we focus on the ways in which sparticles can be produced and detected in experiments. If the superpartners are found, it will confirm that supersymmetry is part of our description of nature. If the superpartners are not observed, it will show that nature is not supersymmetric. We will also discuss the facilities (colliders) where the searches are under way and the future colliders that will provide essential information on the properties of the sparticles.

Whether a particular collider can produce superpartners in a collision depends on the energy and intensity of the colliding beams of particles and on the masses of the superpartners. More energy is needed to produce heavier superpartners. More intensity may be needed to produce enough superpartners to detect a signal. The effective intensity of a beam depends on how many particles are in the beam, how closely they are packed together, and other considerations. For our purposes we can just think of it naïvely: Colliding more intense beams is more likely to produce new particles and will produce more of them once they begin to appear. At the present time there are two colliders that could produce superpartners. One is at the European particle physics laboratory CERN, in Geneva, Switzerland. It is referred to as LEP (from Large Electron–Positron Collider). At LEP, beams of electrons are aimed at beams of positrons so that some collide. The second collider is at Fermi National Accelerator Laboratory, about 40 miles west of Chicago, where a beam of protons collides with a beam of antiprotons. LEP has been increasing its energy every year and has now entered the mass region where Higgs bosons and superpartners are expected to be found, but it can cover only a small part of that region. LEP is now able to operate at an intensity where a run a few months long should give it enough total intensity to see a signal, if superpartners of mass within the limited range accessible to LEP exist. Sometime in 2000 or 2001, LEP will be shut down to make way for a new (proton–proton) collider with considerably higher energy and intensity, the LHC (Large Hadron Collider), also at CERN; LHC is intended to take data beginning in 2005. Fermilab is carrying out an upgrade in intensity (along with a small upgrade in energy) and should begin to take data in 2000. Because we have to probe ever more deeply into nature to gain new insights, the "microscopes" we have to use become larger and more expensive, take longer to build, and require larger groups of scientists. Sometimes people simplistically lament the loss of the golden age when experiments could be done in a few weeks by one or two scientists. But it wasn't always that way. At the beginning of the twentieth century, Marie Curie stirred huge, hot, smelly, dangerous pots filled with pitchblende to extract a tiny bit of radium, for about as many years as it took to upgrade Fermilab to hopefully produce some superpartners at the beginning of the twenty-first.

In a collision in which energy is converted into massive particles, the number of new ones produced could be one, two, or more, depending on the properties of the particles. Because the new particles often carry some charge, and the total amount of charge in a process cannot change, and the initial pure energy carries no charge of any kind, usually two new particles are produced (with opposite signs of their charges, so the net charge produced is zero). Thus only particles of mass about half of the total available energy, or less, could typically be produced.

The available energy is not the only consideration. The rate for producing each new particle depends on the particle's mass and on details of how it interacts. Suppose the collider had beams of sufficient intensity to produce one new particle event a year, on average. In a given year there might or might not be an event, and even if there were, with only one event we might not be convinced that something new had really occurred, because detector malfunctions or an event coincident with a cosmic ray collision or some odd thing could perhaps fake an event. Depending on how common fakes could be, we might not be confident of a signal until ten, twenty, or some larger number of events had occurred. So a signal might be found by just making the collider beams have higher intensity, without an increase in energy. That is basically the approach being used at Fermilab.

Now let us consider the Fermilab collider. At Fermilab, protons and antiprotons are accelerated to energies about ten times larger than the energies the electrons and positrons get at LEP and are then caused to collide. Antiprotons don't occur naturally in our matter world—they run into protons and annihilate very quickly. At Fermilab, lots of antiprotons are produced by hitting nuclei with protons, and then, before they can annihilate, they are separated from protons by appropriate use of electric and magnetic fields and are stored in a chamber from which as much air as possible has been evacuated so that there are as few protons as possible to annihilate with. When enough antiprotons have been collected, they are accelerated and collided with the protons.

At an electron–positron collider such as LEP, all of the energy of the electron and positron is available for conversion into the mass of new particles. For the proton–antiproton collisions like Fermilab's, part of the energy is not available. That's because protons really should be thought of as several quarks bound by a few gluons, and the collisions that produce a lot of energy to convert into mass are those of the point-like quarks and gluons. The quarks and gluons move around inside the proton. At any given moment, it is most likely that the energy of the proton is shared by all of the quarks and gluons more or less equally, so the ones that actually collide are carrying only some fraction (such as a fifth or a tenth) of the proton's energy. Thus the most typical collisions at Fermilab have about as much energy available as those at the lower-energy LEP. But fortunately, some of the collisions have up to nearly ten times the LEP collision energy. The challenge is to have so much intensity that there are a lot of the collisions with a lot of energy available for conversion into mass. Fermilab has been undergoing an intensity upgrade to achieve that goal. Before the upgrade, the intensity at Fermilab was too small to see a signal for most superpartners. At the same time, the detectors are being upgraded so that they can handle an increased event rate and also can recognize events with superpartners better. The first data from the upgraded collider and detectors could come in 2000 if no unanticipated obstacles arise. Results don't come immediately—it takes time to learn how to get to the maximum intensity the collider is capable of, and it takes time for the physicists to learn how to interpret information from the detector. Sometime in 2001, it may be possible to begin to see signals of superpartners if they are there, though it could take longer.

RECOGNIZING SUPERPARTNERS

Recognizing the presence of superpartners may not be easy. It's not that we don't know how they will appear. Supersymmetry is a very well-defined and powerful theory. In fact, the main thing that we don't know about it is whether it actually describes nature or is just a nice idea. For practical purposes, we also need to know the masses of the superpartners to predict their decay patterns, but given those, everything else is determined. If we could calculate or guess the masses, then we could calculate the number of superpartners that would be produced at LEP and Fermilab and how they would appear in the detectors. Because we don't yet know the masses, we simply do calculations for a variety of assumptions about the masses. Then we examine the results to see what typical patterns of final particles are likely to appear in the detectors, and we plan to look for those. Several kinds of problems arise. One is practical. If you want to identify birds in a tree, you will do much better with very good binoculars than with poor ones. Similarly, you can tell different particles apart better with a higher-quality detector. But that costs more, and the funding for detectors is very limited; at present, it is just barely enough. Sometimes the technology necessary to have the detector do what is needed doesn't yet exist. And sometimes different people collaborating on the detector have different physics goals—for example, having the detector more sensitive to one kind of physics may decrease its sensitivity to another if total funding is fixed. If signals for superpartners seem to be in the data, it will probably be possible to get more funds and improve the detectors to confirm the signals and identify their implications.

The second problem is different. Standard Model processes lead to the production of heavy particles too, W's and Z's and top quarks. Sometimes these can mimic the new signatures expected for supersymmetry. These events are called "backgrounds" for the new physics. As is true of cameras and people, the images from detectors are often a bit fuzzy. With perfect images it would be possible to tell which was which, but in practice it is harder. Below we will look at some examples of possible signatures and some of the associated backgrounds. These problems are serious. For example, in 1995 an event was detected at Fermilab that was one type expected if superpartners were being produced. One event of an otherwise unlikely kind is exciting, but before it is considered a discovery, either more such events must occur, or we must observe some other type of event that should occur if the first does. After some study, we were able to identify a different kind of event that should occur if the first one was indeed production of a superpartner, with no Standard Model backgrounds, so if events of that kind were found, it would probably confirm the discovery. Unfortunately, one of the two detectors at Fermilab lacked the components needed to separate this kind of event from others. The other detector could identify such events, but in that detector, such events could also be faked by other processes in the detector. The experimenters actually found lots of candidate events, but when they estimated the number of fakes, it came out to be almost as large—too close to the observed number for anyone to be confident of a signal. During the next run at Fermilab, with improved detectors and much more intensity, this issue will be easily settled. Other kinds of background problems will arise for other possible signals. The ability to find the superpartners if they are there will depend strongly on the quality of the detectors.

A third problem is intrinsic to supersymmetry. In any collider process, two superpartners will be produced. Each either is the lightest superpartner or will decay into the lightest superpartner and other particles. (In some cases, the decays can occur after the superpartner has exited the detector, but we'll ignore complications like that.) Thus there will be two lightest superpartners in every event. Each of them can carry off energy in any direction. Always before, when previously unknown particles were produced at colliders, it has been possible to find a unique, directly visible signature of the new particles, with some of the decay products or combinations of them having a unique energy that stood out from any possible background. In the supersymmetry case, that can never happen because of the two missing lightest superpartners, so it will be much harder to convince anyone who is not an expert that a real signal has been observed. On the other hand, because this extra energy is carried away, compared to what happens in nonsupersymmetric processes, one of the main ways to search for supersymmetry is to build detectors that are very good at detecting all of the energy of the Standard Model particles and summing it up. This possible "missing energy" effect will be the first thing searched for at every new collider and collider upgrade. To say it a little differently, at LEP, Fermilab, and future colliders, if superpartners are produced, two of them will be produced in each event, and they will decay immediately, so the only direct indication of their presence will be the missing energy. (This is illustrated in Figure 5.5.)

There is another fascinating issue that has a lot to do with making detectors expensive and difficult to build if the goal of the detector is to "see" new physics. Because collisions have been studied for over half a century, nearly all of the collisions produce familiar and well-described results. Only a few events—fewer than one in ten billion—are expected to contain superpartners! If the detector records all the events, it becomes effectively impossible to pick out the superpartner ones. For example, if we could examine one event a second to decide whether it contained superpartners, it would probably take over a thousand years until the first one was found. The detector must therefore be designed to make decisions in billionths of a second about whether to retain an event as one with a candidate superpartner (this is called "triggering"). If it does this wrong, it will miss superpartners even if they are there. Building a detector that makes new discoveries requires talented people.

Once a signal is seen, we do understand how to become confident that it really is a signal. First, people carefully calculate the Standard Model backgrounds and demonstrate one by one that none of them can be misinterpreted as such a signal. Second, it is necessary to demonstrate that the observed production rates and different decay modes and masses of the superpartners are consistent with what is expected. That is a very strong constraint. The apparent production rate for a proposed signal for superpartners could be much larger than the expected rate, in which case we would know it was not a real signal; it would be expected to be either a technical error in the analysis or a statistical fluke that would go away as more data were gathered. Or the "observed" decay might be a minor one, implying that the major decay must also be present, but careful scrutiny might fail to reveal the major decay in the data; this too would imply that the signal was not real. If a signal passes all such tests, it can be taken very seriously. Finally, once one sparticle signal is found, others have to be nearby. They can be looked for in different signatures. This part of the study is less certain initially, because what is expected can depend on the unknown masses, but once some masses are measured, additional ones can be estimated. These kinds of tests are possible here because supersymmetry is such a well-defined and well-understood theory. In many areas of science it is harder to know what is expected, but here, once the relevant masses are known from the tentative experimental result, the other predictions can be calculated.

Sparticles: Their Personalities,
Backgrounds, and Signatures

Those of us who think about sparticles daily form images of their behavior. We like some more than others. Selectrons and smuons, interacting with photons and Z's, decay into electrons or muons plus the lightest superpartner, and have no interactions via the strong force. But production of a W pair can look a lot like selectrons or smuons unless one looks very closely. Winos and charged higgsinos do a complicated thing: When the Higgs mechanism operates they can't be told apart, so what we see will be a mixture, called charginos. Similarly, photinos and Zinos and neutral higgsinos can't be told apart, and we will observe mixtures called neutralinos (see Appendix C if you want to know more about these mixtures). Gluinos have the strongest interactions of all the superpartners, so if they are not too heavy, more of them will be produced at Fermilab than any other sparticle.

In this section we will look at several examples of how sparticles might be produced at LEP and Fermilab. It's probably clearest to show the Feynman diagrams for their production and decay in the same diagram. The initial state is the beams that collide: electron (e) and positron ($\bar{e}$) for LEP, quarks (q) or gluons (g) inside protons and antiprotons for Fermilab. The diagrams are read with time moving from left to right. All the diagrams are constructed by combining vertices such as those of Chapter 2 plus the additional vertices that make the theory supersymmetric, as in Figure 4.1. First the sparticles are produced, and then they decay into the final products. The particles on the right sides of the diagrams actually are observed in the detector. Neutrinos and the lightest superpartner are observed in the passive sense that they carry away energy even though they are not seen directly. In the diagrams the lightest superpartner is denoted by LSP, because we don't yet know which superpartner is the lightest. Often in the figures and text below, to simplify the description, we won't distinguish between the particles' charges, or between particle and antiparticle, if it doesn't matter for explaining what happens. Sometimes the detectors also can't distinguish between particles and antiparticles. (Readers who don't want to focus on the ways in which superpartners might be observed can skim the following until after the paragraph discussing Figure 5.6.)

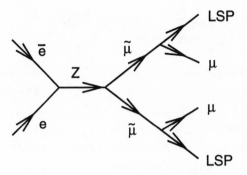

FIGURE 5.1. A possible way to produce smuons at LEP. This diagram can be read by following the arrows. The particles on the right, muons and the two lightest superpartners, emerge from the collision into the detector. The muons will be observed, and the two lightest superpartners will escape the detector, carrying energy off.

Perhaps smuons ($\tilde{\mu}$) and selectrons ($\tilde{e}$) will be light enough to produce at LEP. Figure 5.1 shows how smuons might be produced. The electron and positron beams collide to create a Z boson, which turns into a smuon pair. Each smuon then decays into a muon and the lightest superpartner. In the detector, this would be seen as an event with two muons (μ) moving off in arbitrary directions and missing energy carried off by the LSP's.

There would be no problem in identifying such events in the detector—muons can be identified and their energy measured easily, so it would be simple to learn whether the energies of the two muons added up to significantly less than the sum of the electron and positron energies. But there is a significant background that could fake such events, because a W boson can decay into a muon (μ) and a neutrino (ν). Thus, as shown in Figure 5.2, a pair of W's will be produced as a normal Standard Model process even if there are no superpartners, and each W will decay, sometimes into a muon and a neutrino. The neutrinos escape the detector, so again the event looks superficially like smuon production. It is still possible to find a signal if it is there, but only by having enough intensity to produce a number of events and applying considerable analysis. First, one can calculate the expected number of WW events. If the observed number is larger than the number expected in the Standard Model by an amount beyond what might occur as a consequence of statistical fluctuations, that is evidence for smuon production. Second, the typical directions of the muons, and the typical values of the missing energy, will differ somewhat in the two cases, so if there are enough events, one can see patterns that differ.

Another interesting possibility for how the first superpartners might be discovered at LEP is selectron production in the special case where the lightest superpartner turns out to be essentially a higgsino, the superpartner of the Higgs boson. For technical reasons, selectrons ($\tilde{e}$) mainly decay into electron and photino ($\tilde{\gamma}$) but not into electron and higgsino ($\tilde{h}$), so the events would give a selectron–antiselectron pair, which then would decay into an electron plus a positron plus two photinos. But the photinos would then quickly decay into the lighter higgsinos plus a photon, so the detector would observe an electron, a positron, two photons, and the missing energy carried off by the two higgsinos, as in Figure 5.3 on the next page.

Such events have very small backgrounds (that is, Standard Model events that could mimic the supersymmetric event), and in this case the behavior of the background events in terms of the energies and directions of the photons is rather different from those of the signal, so if more than a few such events occur, it will be possible to separate them from the backgrounds.

LEP will take data in 2000. The lightest superpartners are likely to be selectrons, smuons, staus, stops, charginos, and neutralinos. If any of them is in the LEP range, they will be produced there and presumably

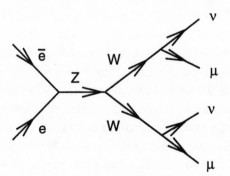

FIGURE 5.2.   This process provides a significant background that could be mistaken for the smuon production of Figure 5.1, because the final state looks very similar in the detector. The neutrinos also escape the detector, carrying off energy.

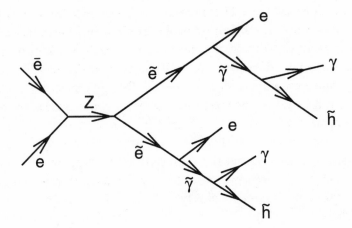

FIGURE 5.3 A possible selectron production event at LEP for the case
where the LSP is a higgsino.

detected. If not, we will learn that the sparticles are too heavy to be found
at LEP. We could go so far as to question whether the superpartners exist
if they are not found at LEP, but the reach of LEP is not sufficient to support that conclusion. We will see in Chapter 7 that LEP also has a significant opportunity to detect a Higgs boson.

Starting sometime in 2000, the Fermilab collider will take data with several times the useful energy of LEP and several times the intensity, so its
reach for superpartners is quite a bit larger than that of LEP. In addition,
because the LEP electron and positron beams do not have strong interactions, they do not produce gluinos, whereas the Fermilab proton and antiproton beams contain gluons and quarks and can produce gluinos with
large probabilities. There are significant theoretical arguments (but not
proofs) that gluinos should be light enough to produce at Fermilab, so this
could be a significant advantage for Fermilab over LEP. At Fermilab, in
general more reactions can occur than at LEP, so the backgrounds that
could obscure signals are typically harder to eliminate there, but with experimenters strongly motivated to make discoveries, it will be possible.
One process that could occur at Fermilab is just like that of Figure 5.3,
with the initial electron and positron replaced by an up quark from a proton and its antiparticle from an antiproton. The final state has an electron,
a positron, two photons, and missing energy. The Fermilab detectors can
identify such a final state efficiently, and it is one that is hard to fake, with

very little background. As we noted earlier in this chapter, one event of this type has been reported at Fermilab, and analysis of its properties showed, surprisingly, that it passed all of the consistency checks to be a real candidate for sparticle production. The two detector groups will be eagerly watching for events of this type after the upgrade.

Figures 5.4 and 5.6 show two more of the many processes that could occur at Fermilab.

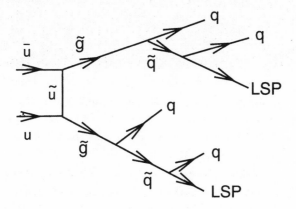

FIGURE 5.4   A process by which gluinos might be produced at Fermilab. They decay through an intermediate squark stage. The final state has four observable quarks or antiquarks, and two lightest superpartners that carry off energy.

Figure 5.4 illustrates one way to produce gluinos ($\tilde{g}$), starting from quarks in the colliding protons. The gluinos quickly decay into squark ($\tilde{q}$) and quark, and the squark quickly decays into quark and lightest superpartner (LSP). Ignoring the distinction between quarks and antiquarks, the final state has four quarks and the escaping lightest superpartners. Because quarks and gluons interact strongly, they produce several particles in the detector, and they are moving energetically, so they effectively appear in the detector as "jets" of particles. Lots of Standard Model processes also give final quarks, but usually with much less missing energy because they produce none of the LSPs that carry energy off. Investigators will recognize a process such as this one by looking for extra events with several jets and considerable missing energy. In practice, of course, the amount of missing energy is studied quantitatively, perhaps

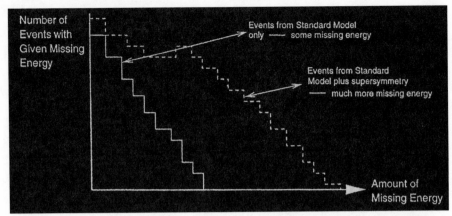

FIGURE 5.5.   This figure shows how it might be possible to learn that the process shown in Figure 5.4 is occurring even though individual events look very much like events that would occur even if superpartners were not being produced. Events with superpartners are expected to have more undetected energy because of the escaping LSP's, so experimenters make a graph showing the number of events with a given amount of missing energy. If superpartners are not there, the result should look like the solid line labeled Standard Model, whereas if superpartners are being produced, the result should look like the dashed line labeled Standard Model plus supersymmetry.

focusing on a graph such as Figure 5.5. If a result such as that of Figure 5.5 is observed at Fermilab, experts will be convinced supersymmetry has been discovered! Convincing the world that an excess of events with missing energy is proof of supersymmetry may be harder.

Figure 5.6 shows how Winos ($\widetilde{W}$) and Zinos ($\widetilde{Z}$) might be produced and detected at Fermilab. (Actually it would be the chargino and neutralino mixtures, but we can ignore that complication for purposes of understanding what basically happens—see Appendix C.) An up quark and a down antiquark, initially in the proton and antiproton, collide by exchanging a down squark ($\widetilde{d}$), producing a Wino that quickly decays into a W and a lightest superpartner (LSP); then the W decays into a muon and a neutrino. The muon is recognized in the detector, and the neutrino and lightest superpartner escape, carrying off energy. At the other vertex, a Zino is produced and quickly decays into a Z and a lightest superpartner, followed by the Z decaying into an electron and a positron. The electron and positron are seen in the detector, and the lightest superpartner es-

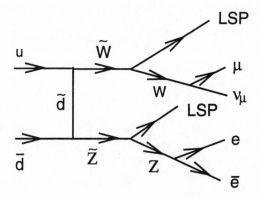

FIGURE 5.6.   A probable process by which the partners of W's and Z's might be produced and observed at Fermilab. The Wino and Zino decay through an intermediate stage of a W and a Z. The final state has three charged leptons (a "trilepton event"), and energy is carried off by two LSP's and a neutrino.

capes. Thus the detector sees a muon, an electron, a positron, missing energy, and none of the strongly interacting jets that are very common. This is a nice event because the leptons can be studied with confidence in the detector, and Standard Model sources of such events are reliably predictable. These "trilepton" events may be one of the best ways to search for a supersymmetry signal at the upgraded Fermilab collider, or to confirm that what is observed is indeed supersymmetry physics if the first evidence comes from another process.

Figure 5.7 shows how a trilepton event of the kind illustrated in Figure 5.6 might look in a detector at Fermilab. It is not an actual event, but a simulation by Jane Nachtman of the CDF detector group at Fermilab. This event assumes essentially that a Wino and a Zino, both of mass about 1.3 times heavier than a Z boson, are produced and that they decay as in Figure 5.6 (except that the muon is replaced by an electron). The top view is a projection of the tracks on a plane, with the colliding beams perpendicular to the page. Lines are reconstructions of tracks of particles and are curved more if they are bent more by the magnetic field of the detector; the more energetic the particle, the less it is bent. The arrow shows the direction and size of the missing energy carried away by the two LSP's. (In a real event, this would be the amount of energy missing when the energies of all the observed particles were added.) The boxes around the outside show that

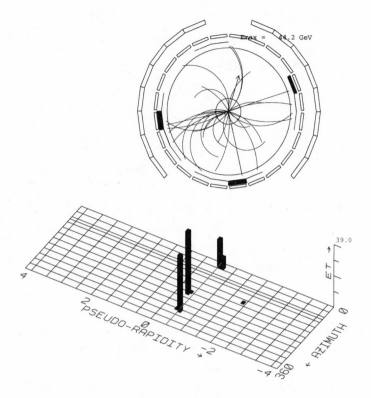

FIGURE 5.7.   A simulated "trilepton" event that might be a signal of supersymmetry (see text).

electrons deposited a lot of energy in those detector elements. The lower view is as though the cylindrical detector were unrolled to be flat; it shows the energy deposited by the three electrons. From these computer reconstructions, it is possible to deduce the properties of the event and study whether it is likely to be recognized as a signal of the production of particles other than Standard Model ones.

Several arguments imply that some sparticles are within the reach of Fermilab. The strongest is based on the explanation supersymmetry gives for the Higgs mechanism of the Standard Model, as described in the last chapter. Basically, the argument is that because supersymmetry provides the mechanism that accounts for the masses of W and Z, the sparticle masses cannot be much heavier than the W and Z masses themselves. Fermilab has already produced and detected thousands of W's and Z's. When this argument is framed in a technical form, it implies that gluinos and probably charginos and neutralinos and stops should be in the Fermilab reach. If they are not, the impressive successes of supersymmetry listed at the beginning of Chapter 4 may be meaningless coincidences.

Suppose superpartners are indeed detected at Fermilab. What then? There will be great excitement, because we have dramatically increased our understanding of how nature works and, most important, because we have further opened the window to the ultimate, or primary, theory. Just knowing that nature is indeed described by supersymmetry tells us that we are working in the right direction to make more progress. Supersymmetry may be the last stage on the path to the primary theory that can be studied experimentally.

Then the fascinating work begins. The properties of the superpartners can tell us an immense amount about how the primary theory works. In trying to formulate the primary theory, we can imagine many alternative approaches, and each approach will lead to superpartners with different masses and interactions. Thus it will be very important to take detailed data to deduce the needed properties. It will not be easy; each process in Figures 5.1–5.6 depends on a number of superpartner properties. To determine each mass and interaction, measurements will have to be made on many processes. It will be a challenging task, but an exciting one.

VISIT FERMILAB

If superpartners and Higgs bosons are observed and studied at Fermilab, it will be a wonderful scientific success. But Fermilab has another distinction—it is not only scientifically but also visually and esthetically an exciting place to visit. Robert Wilson, the director of the laboratory in charge of its design and construction (in the late 1960s), and of its first years of productivity, was a sculptor as well as a fine physicist, and he made the lab beautiful (not a description usually applied to government or scientific facilities), as is illustrated in the accompanying photos. He even convinced Congress and the Department of Energy to include a 1% allocation for esthetics in the construction budget. Recently Dan Goldin, the head NASA administrator, visited Fermilab and was told about the allocation for esthetics. He was reported as saying, "You couldn't do [that] in the government today."

The main "high rise" building, now Wilson Hall, houses offices and many activities of the lab; it is shown from the outside in the upper-left picture on the first page of Fermilab photos (pages 90–91), from the inside

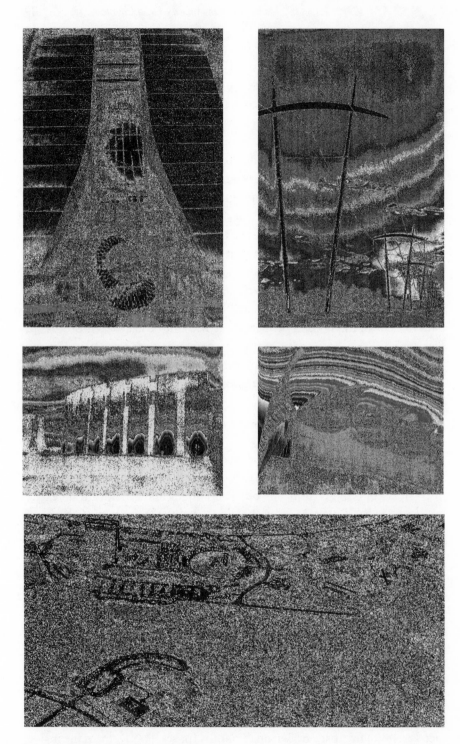

Views of Fermilab (Fermilab photos).

Views of Fermilab (Fermilab photos).

in the middle-left picture on the second page, and in an aerial view at the bottom of the first page. The other photos show various structures and buildings around the lab.

Visitors to Fermilab can see the accelerator, the four-mile main ring, the detectors, and the site in general. There is a fine view from the top of Wilson Hall. Tours can be arranged; self-guided tours are possible if the detectors and the accelerator are not included. There is a cafeteria and even a gourmet restaurant (currently serving Wednesday lunch and Thursday dinner) run by the spouse of one of the experimenters. Information can be found at www.fnal.gov. Fermilab is a fascinating place to visit.

FUTURE COLLIDERS

Although the first signals of supersymmetry should be found soon, with luck at LEP and probably at Fermilab, if present ideas are basically right, for several reasons these colliders will not be sufficient to untangle fully the information needed to formulate the primary theory. Fermilab should be able to make many measurements over a decade or more that will be important in this process. But we know enough about the theory to understand already that Fermilab data alone will not be enough. First, some of the superpartners and Higgs bosons will be so heavy that they will not be produced in detectable numbers at Fermilab. The CERN LHC, with about seven times the energy and ten or more times the intensity of Fermilab, will easily be able to study these heavier ones and obtain essential information. Too few superpartners will be detected at Fermilab to settle some issues. Recently we have been able to demonstrate that some essential measurements can be made only if at least one new collider beyond LHC is constructed: a lepton (electron–positron or muon–antimuon) collider with the extra capability of selecting the spin projection of one beam (the jargon is "polarized beam"). That collider is needed because a full characterization of the superpartner properties requires at least one measurement for each unknown parameter, and we can show that to obtain enough measurements, it is necessary to measure production rates and distributions with differently polarized beams. Of course, that collider must have enough energy to produce some superpartners, though not necessarily all.

One such collider has been under study for over a decade, at the Stanford Linear Accelerator Center in Palo Alto, California, and also in Japan and

Germany and at CERN. We don't know yet where it would be located if it were built or which specific technology would be chosen to accelerate the beams. In 1987 I called it the NLC, for Next Linear Collider, and the name stuck, but once a definite design and location are chosen (assuming that happens), it will get a new name. A decade of research and development on NLC design has demonstrated that the NLC could be successfully built and that it would offer sufficient intensity to produce superpartners in its energy range if they were there. The remaining challenges are to bring the cost into an affordable domain and to negotiate the complex funding issues. An NLC that could produce some of the lighter sparticles would complement Fermilab and LHC very nicely so that together they could provide the data that would enable us to deduce all the relevant properties of superpartners.

More recently, the idea of another kind of lepton collider has emerged. Muons decay about a millionth of a second after they are created, so you might think it wouldn't make sense to try to collide them. But one of the well-tested predictions of special relativity is that moving clocks tick more slowly than stationary ones, and an unstable particle is like a clock. Moving muons live longer. At the energies needed to make an interesting collider, the muons live long enough so that one can make beams of them and obtain large collision intensities. More study of the technical challenges that have to be overcome to make a muon collider useful for physics is needed before we can say whether it is possible. In the long run, this may be a practical way to produce heavier superpartners with polarized beams. The favored acronym here is FMC, for First Muon Collider; its energy and location have not yet been determined.

Because colliders are always at the frontier of what is known, existing techniques and technologies must always be extended to design and construct them. It necessarily takes a decade or so to be sure they will work as intended and to find relatively inexpensive ways to construct them. The construction itself takes six to ten years. Thus, unfortunately, we can be sure that new devices will not suddenly emerge and give us unexpected ways to produce new particles and study them. LHC is already under construction. It uses some technologies that are at the edge of what has been done, so its completion may take a little extra time, and it is marginally underfunded. Perhaps it will make the planned turn-on date of 2005, but it could be rather late. NLC has been under study since the late 1980s. It could be approved for construction as early as 2002, and it could take data before 2010 if the process went ahead smoothly. A muon collider is a newer idea, and more R&D is needed before we can tell whether it could work in a scientifically useful way, so it will take longer. Any newer technologies will have to meet the combined requirements of high energy and high intensity in order to be useful for new-particle physics, so they are for the more distant future. Table 5.1 shows a summary of existing and planned colliders worldwide that could produce superpartners.

TABLE 5.1. Colliders Present and Future.

Collider	Particles Colliding	Total Energy	Location	Status
LEP	electron, positron	200	CERN, Geneva	will shut down in 2000 or 2001
Fermilab	proton, antiproton	2000	Chicago	will begin running in 2000 after upgrade completed
LHC	proton, proton	14,000	CERN	initial running planned for 2005
NLC	electron, positron	500–1500?	?	full technical proposals being written
FMC	muon, antimuon	?	?	RIP

This table summarizes some features of existing and planned colliders that could produce superpartners or Higgs bosons. An official proposal will be made for NLC in the U.S. in 2001 or 2002, after 13–14 years of R&D; additional proposals may be made elsewhere in the world. Selection of an NLC site somewhere in the world will occur during the process of obtaining and funding to build it, assuming it is indeed funded. Construction will take about seven years. R&D on a muon collider has been underway for several years. NLC can naturally have a polarized beam as needed to make measurements essential to understand supersymmetry, as can a muon collider under certain circumstances. The units for the total energies are gigavolts, but the reader only needs to look at relative energies, keeping in mind that for the colliders involving protons not all of the energy is useful for producing new particles.

CAN WE DO THE EXPERIMENTS WE NEED TO DO?

Will we need colliders that provide even larger energy or intensity than those in Table 5.1? Can such colliders be built? One sort of limit could be financial. Facilities that can probe nature more deeply than ever before, so we can continue the quest to understand the physical universe, have to be larger and more complicated than previous ones. They necessarily involve new techniques; the older facilities would have been extended if that were possible and if it were likely to provide new knowledge. New colliders will be expensive. Only governments or the largest foundations can build them. In fact, society always comes out ahead, even from a purely financial perspective, when it builds such facilities, because new developments lead to "spinoffs" that in turn lead to multibillion-dollar industries. A recent example is the World Wide Web, introduced by particle physicists to confront new challenges involving analysis programs for detectors and to handle unprecedented amounts of data. There are many more examples. Regrettably, the leaders who decide on the funding may not understand that it is an investment, especially in view of the fact that the returns on the investment are not likely to occur within their term of office. Society should be funding every scientifically and technically justified experiment. Experience shows that it is not a zero-sum battle; every part of society wins if these projects go ahead.

Another limit could come from a lack of "human capital." Significant numbers of talented and committed people are needed to build a state-of-the-art facility and make it work. It's not possible to start from zero. The people who work at existing facilities are experienced and "up to speed"—they know how to answer major scientific questions with new facilities. If progress is blocked, these people disperse to other jobs, and many leave research altogether. It is not possible to reassemble them. Society loses not only because the new developments do not occur but also because talented people are forced out of research. If the United States does not support construction of the required facilities, the associated research fields here will die. Some of the best researchers will work elsewhere, and some will leave research. The remaining personnel, hamstrung by loss of the elite among them and working in facilities stretched to their limit, are less likely to do the job right.

Whether we will need more powerful colliders depends on whether we find the sparticles and the Higgs bosons (Chapter 7) in the coming years and on whether there are surprises in what is found at LEP, Fermilab, and LHC. One can imagine a world in which all of the evidence about the primary theory that could ever be found at colliders can be found at Fermilab and LHC and NLC, but we won't know whether the world is that way until we get there. That one can imagine such a world is a very strong statement that could never have been made before in history, because now we have the Standard Model and we have supersymmetric approaches that can provide a window to the Planck scale and the primary theory, and help with the study of the unanswered questions about the universe (more on this in Chapter 9). (In addition to our need to learn about the new particles and interactions that can be studied at colliders, we need information about neutrino masses, dark matter, the stability of the proton, which interactions are invariant under time reversal [the so-called CP violation], and more. The experiments that can tell us about these issues are also challenging, but they are possible.)

Even if we need more energetic or intense colliders, we may not be able to have them. The only way to increase the energies of the particles is with electric fields, and there are practical limits to how strong an electric field we can make in a situation that could accelerate particles. The higher energies imply that the length of the acceleration part of the collider must be longer if it is a linear one—eventually so long that it is too expensive or impractical. If magnets are used to curve the path of the particles into a circular one, limits on the strength of the magnetic fields of real robust magnets that have to work for years impose a limit on how energetic the particles can be. The probabilities of producing previously unknown particles or of finding previously unknown interactions necessarily decrease at higher energies (that's not obvious, but it's a consequence of quantum theory), so higher-energy facilities must have higher intensity too, and many innovative techniques to accelerate particles to higher energies provide only low intensities, so they are not useful for particle physics. We are not yet at the technological and financial limits of what we can do to probe nature more deeply via colliders, but those limits are drawing near.

One of the exciting aspects of the view of the world suggested by supersymmetry is that we may be able to get the data needed to formulate and test the primary theory from the facilities already under construction and those already being designed. The masses and production rates of the superpartners will tell us a great deal about the form the primary theory takes, as will the nature of the dark matter of the universe, and as will possible tiny effects in the way neutrons and electrons behave in electric fields, and whether the proton is a stable particle or eventually decays, and more. We won't know until we succeed whether we really can have all the data we need to formulate and test the primary theory. But for the first time in history, it is now reasonable to argue that we might do that within a couple of decades.

What Is the Universe Made Of?

By the early 1980s, people recognized a surprising implication of a super-symmetric world with a stable lightest superpartner. It turned out that in a supersymmetric world, after the Big Bang the universe would contain many of the lightest superpartners—so many that their gravitational attraction for the other particles would have significantly slowed the expansion rate of the universe. The Big Bang was the creation of huge numbers of all of the basic particles: quarks and leptons and W's and Z's and photons and gluons, and also superpartners. The particles moved throughout the expanding universe (which was then small, perhaps the size of a soccer ball) colliding with each other. The heavy superpartners were decaying into the lighter ones and Standard Model particles. After a while, only stable particles were left—the up and down quarks, gluons, photons, electrons, neutrinos, and the lightest superpartners. We speak of all these particles as "relics" because they are all left over from the Big Bang itself, or from decays and annihilations of other particles into them soon after the Big Bang.

Eventually the quarks and gluons bound into neutrons and protons, the neutrons and protons bound into the lightest nuclei (deuterium, lithium, and the very stable helium). Joining with electrons, the nuclei formed atoms. One of the great successes of our understanding of the first few minutes of the universe is the prediction and subsequent confirmation that about 24 percent of the atoms in the universe are helium. Eventually, clumps of hydrogen and helium condensed, from mutual gravitational attraction, and formed stars. Some of the atoms remain as "dust" spread throughout the universe. From observations of starlight and x-rays, we can deduce that a little over a tenth of the matter in the universe is in stars and dust.

How do we learn how much total matter there is in the universe? Basically, astronomers observe the motion of stars in galaxies, of galaxies in clusters of galaxies, and so on. All of them move in paths described by Newton's laws of motion and gravitation. The strength of the gravitational force on a star is proportional to the mass of whatever is attracting the star, and it decreases if the star is farther away. In the solar system, for example, this means that planets orbit the sun in longer "years" if they are farther from the sun. One would expect a similar result for stars moving in galaxies, but it turns out that their speeds are much greater than expected. Analysis shows that rather than most of the mass of a galaxy being at the center, as it appears to be from the starlight we see, the mass is spread throughout the galaxy more or less uniformly.

Another approach to estimating the total matter content is to compare the recent expansion rate of the universe with the rate much earlier. If there is lots of mass, the recent rate will be slower than if there is only a little mass, because of the gravitational attraction of the additional mass. One tool used to make these kinds of measurements is the "Doppler shift" of the light emitted by atoms. Every hydrogen atom in the universe emits light composed of a unique set of colors (the color of light is determined by its wavelength, or, equivalently, its oscillation frequency, so any of these words could be used to describe these effects). Similarly, every carbon atom in the universe emits light of a different unique set of colors. A familiar experience is the way sound is shifted to a higher pitch (frequency) if the source of sound (such as an ambulance siren) approaches us; the shift is larger if the speed of the approaching siren is larger. If the siren is receding from us, the shift is the other way, to lower frequency. For light the same effect occurs. An approaching source appears to be higher in frequency, or bluer, than if it were stationary, and a receding source appears to be redder—it is red-shifted. The original discovery of the expanding universe came, in the 1920s, from observing that the light from distant stars is red-shifted. The amount the light is red-shifted at a given distance gives us information about how fast the universe is expanding.

A third approach to learning how much energy and matter there is in the universe is to study the energies of the photons left over from early in the history of the universe; their energies are affected by the other sources of gravity. Combining various kinds of information like these enables astronomers and cosmologists to deduce the total amount of matter in the universe.

WHAT PARTICLES ARE THERE IN THE UNIVERSE?

There are lots of photons in the universe, about 400 in every cubic centimeter. About 1 percent of the static on the screen of your TV when no channel is being received is due to those photons. The photons can have a range of energies. Another of the great predictions of the Big Bang picture that gives us confidence that we understand it was the calculation, and then observation in the 1960s, of the number of photons of each energy. Because energy has gravitational effects, these photons also affect the expansion rate of the universe, but their effect on the expansion rate turns out to be very small.

Although the relic neutrinos have not yet been directly observed, the number of them can be calculated by the same methods that produce the right numbers of photons and of helium nuclei, so we can have confidence in the result. There are nearly as many neutrinos as photons. To deduce their effect on the expansion of the universe, we need to know not only how many there are but also their mass. We expect neutrinos to have mass, because theories that extend the Standard Model normally predict that they do, but there is as yet no compelling estimate of their mass. Observations so far imply that neutrinos have mass, but these observations give only a lower limit of the mass, not a measurement. What is actually measured is the difference between two neutrino masses, and that is not zero, but the separate masses have not yet been determined. An interesting astronomical argument implies that at most about 20 percent of the matter in the universe could be neutrinos. That is because neutrinos are so light, and interact so weakly, that even if all the matter in the universe were neutrinos, they would just keep moving and never clump into galaxies as is observed. Altogether, current reasoning suggests that neutrinos do indeed make up 10 to 20 percent of the matter in the universe, but the true result could be less. Experiments now taking data or under construction could settle this issue over the next decade.

Combining all these forms of matter, including stars and dust, we see that at most about a fifth of the matter in the universe is accounted for. Whatever the rest is, it is not made of the particles that we are made of or even of those we know about! In the early 1980s this accounting had not yet been done, and it was not recognized that some then unknown form of matter was needed to explain the total amount of matter. Once investigators realized that the lightest superpartner would provide a form of matter that would affect the expansion rate of the universe, estimates were made of how much supersymmetric matter there might be and what it would do to the expansion rate. This was already a significant test of the possible validity of supersymmetry, because it could have happened that for any reasonable lightest superpartner mass, the universe would be "overclosed"—that is, the lightest superpartner would exert so much gravitational attraction that the universe would stop expanding and collapse back long before it got to be its present age. You are reading this, so that didn't happen. In fact, it turned out that the lightest superpartners would give about the right amount of matter to account for the total amount, though until the mass and the interaction properties of the lightest superpartner are measured, it will not be possible to confirm this.

Because it interacts too weakly, superpartner matter would not join in the reactions that occur in stars to make starlight. The superpartners do interact with each other and with Standard Model particles a little, and they do cluster with other matter loosely into galaxies because all matter feels the gravitational force, but if they were the only particles that existed, the universe would be dark, just as it would be with neutrinos, so these forms of matter are called *dark matter*. The lightest superpartner is expected to have a mass about that of the W mass (very heavy compared to protons and light nuclei), so essentially all of its energy will be in mass rather than motion. We know that higher temperatures correspond to faster motion of particles, and conversely, so we speak of the lightest superpartners as *cold dark matter*. Neutrinos will be much lighter than the lightest superpartners, so much more of their energy will be in motion; thus they are called *hot dark matter*.

Sometimes people who are not familiar with the history assume that proponents of supersymmetry introduced the lightest superpartner, or even the idea of supersymmetry, to account for the dark matter. On the contrary, the idea of superpartners and that of a stable lightest superpartner emerged without awareness of a connection to dark matter. The prediction that they would form cold dark matter was made without knowledge of the need for a new kind of cold dark matter. When the prediction was made, it was known that there was more matter in the universe than could be accounted for by stars and dust, so something like dark matter was needed. But it was thought that the "missing" matter could be ordinary nuclei (consisting of protons and neutrons) in big planets such as Jupiter, or white dwarf stars that no longer were visible, or that some of it could be neutrinos. Today we know that none of these can account for the missing matter—that is, if we add up all of the forms of matter predicted by the Standard Model, we cannot explain the observed amount of matter in the universe. The lightest superpartner could account for it. Even if supersymmetry provided only the right stuff to account for the majority of matter in the universe, it would be very important. It can do that, and it can account for the Higgs mechanism, and so much more as well (as described in Chapter 4 and elaborated further in the following chapters), so of course many physicists are very excited about it. At the same time, it is good to keep in mind that this is RIP, so new data and new ideas can change our understanding. For example, other possible forms of matter that people have speculated about could perhaps also account for the dark matter.

Is the Lightest Superpartner the Cold Dark Matter of the Universe?

How can we find out whether the lightest superpartner (LSP) is indeed the cold dark matter of the universe? To be certain, we will need to do two things: Observe the lightest superpartner in laboratory experiments, and then measure its properties well enough to calculate how many lightest superpartners are left from the Big Bang, allowing for depletion by those that annihilated as the universe evolved. This section is a little technical, so don't be concerned if some aspects seem complicated. In Chapter 5, we surveyed ways in which superpartners could be detected at colliders. Superpartner events have two lightest superpartners escaping the detector. Detection of the lightest superpartner at colliders involves indirect reasoning rather than direct observation, but it is only one more step in the chain of reasoning going from the photons bouncing off objects into our eyes to analysis of digitized information from detector modules. If the lightest superpartner is the cold dark matter, superpartners will be observed at colliders, and the lightest superpartner will be detected indirectly by these methods. Once such events occur, we can proceed to deduce the properties of the LSP and calculate what fraction of the universe it accounts for.

Actually, there is an interesting subtlety. To be the cold dark matter, the lightest superpartner has to live at least as long as the lifetime of the universe (about twelve billion years), which it does if it is stable. But it is also conceivable that a particle could live long enough to escape from a detector (that takes only somewhat over a billionth of a second) but decay soon after (say in a second or a year) and thus not be the cold dark matter. However, the typical lifetime for a superpartner is much, much shorter than a billionth of a second. Thus if a superpartner escapes from the detector rather than decaying, it is living much, much longer than it would naturally live. This would probably mean that it was indeed stable. But for such an important question as what makes up most of the universe, we want better evidence, so we need to find ways to show experimentally that the LSP's really live as long as the universe and are spread all over the universe today.

There are three methods that people are pursuing to detect explicitly ("explicit detection") the lightest superpartner in a context that confirms its status as cold dark matter. The first, and perhaps the most promising, is "direct detection." If the lightest superpartner is the cold dark matter, there are lots of lightest superpartners just sitting around in the universe. Our sun is moving around in the galaxy, and the earth is moving around the sun. Consequently, all of us, and any detector we build in a lab, are moving through a cloud of lightest superpartners that are more or less at rest in the universe. There might be one or a few in every region the size of a baseball. As we move through them, occasionally (rarely) a nucleus in us or our detector will interact with a lightest superpartner and recoil. Thus it is necessary to make detectors that are very sensitive to such a collision, which is very hard to do. Several approaches are being tried. One is to make the detector in such a way that it is normally a superconductor, which conducts electricity with no resistance or loss of current. Superconductors are extremely sensitive to temperature, and the heat energy they gain from one recoiling nucleus can make them lose their superconductivity, so even the tiny amount of energy from such a collision can have a very large effect that is easily detectable. Making detectors that are sensitive to such small amounts of energy is very challenging. One of the problems is that if a nucleus in the detector decays radioactively, the decay products also deposit energy in the detector, in an amount similar to the collision with the cold dark matter, so it can mimic a cold dark matter collision. If an experimental group reports a tentative signal, there are possible ways to check that it is indeed from cold dark matter. For example, during half of the year, the earth and the sun move in the same direction, so their speeds add, giving a larger velocity and therefore more frequent interactions. During the other half of the year, they move oppositely, so the speeds subtract, giving less frequent interactions; thus the event rate should vary in a predictable way over the course of a year.

A second method of trying to detect the lightest superpartner takes advantage of very different aspects of LSP behavior. The strong gravitational attraction of both the sun and the earth will capture a lot of the lightest superpartners, and they will fall to the center of the sun or the earth. Many lightest superpartners will be concentrated there, so occasionally a pair will annihilate. For example, two lightest superpartners can exchange a stau ($\tilde{\tau}$) and annihilate into a tau (τ) and an antitau as shown in Figure 6.1.

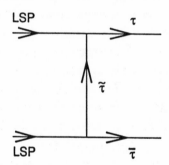

FIGURE 6.1.   Two dark matter LSPs annihilating.

The tau and the antitau in turn will decay, giving a muon neutrino about a fifth of the time. That muon neutrino has an energy that is typically about ten thousand times larger than the energy of a typical neutrino from the sun. Not many such neutrinos will interact, but if one does, it will produce a muon, perhaps in a rock underground. There are detectors now set up, particularly in Italy and Japan, that can observe muons created in the rocks below the detector. Whether cold dark matter composed of the LSP will lead to this chain of events often enough to give an observable signal depends on the mass and properties of the lightest superpartner.

The third method takes the detectors into outer space. Suppose there are cold dark matter LSP's all through the galaxy. Anywhere in the galaxy a pair might annihilate, giving quarks and antiquarks, electrons and positrons, photons and neutrinos in the final state. The normal numbers of electrons and neutrinos and quarks present in space are so large that an excess from lightest superpartner annihilation could not be detected. In outer space, positrons do not encounter an electron so quickly, so they might hang around a while. Antiquarks occasionally combine into an antiproton, which on the earth would annihilate quickly but also might persist in space for a while. There are too many photons for us to detect an excess in general, but an excess might be detectable for photons of a particular energy. Accordingly, investigators have tried to put up detectors that can see an excess of antiprotons or positrons or photons of a particular energy. So far the detectors have all been flown in balloons that stay up hours or a few days; one of the main groups pursuing this approach is headed by my colleague Greg Tarle. They have obtained promising results, and upgraded detectors are being prepared for more flights. An even more ambitious approach is being pursued by a large group headed by Samuel C. C. Ting of M.I.T. and CERN. This group is preparing a large detector that will fly for months on the space station. One of the goals of their experiment is the detection of annihilation products from cold dark matter, with emphasis on the antiprotons.

It's important to understand that none of the explicit detection methods will allow us immediately to conclude scientifically that the particle detected is actually the cold dark matter. To calculate the actual amount of superpartner cold dark matter in the universe, we need to know the number of cold dark matter particles in the universe and their mass. Once there is an observation, the interpretation of the experimental results depends on those quantities, but it also depends on other properties, such as the probability that the lightest superpartner will bounce off nuclei or that it will annihilate (as in Figure 6.1, for example). Without some other way to measure or calculate these other properties we cannot obtain the needed information. Of course, if an effect is observed we will be convinced psychologically, but we will have to do better to really claim we know what constitutes most of the matter of the universe.

It is more likely that we will first observe the lightest superpartners at colliders and that, not much later, one or more of the explicit experiments will confirm that the lightest superpartner truly has lived as long as the universe has existed. From colliders we will eventually be able to deduce all of the properties of the lightest superpartner and its interactions that we need to work out what fraction of the matter in the universe is composed of the lightest superpartner. As I discussed in the previous chapter, that will not be easy; it will probably require data from three colliders (Fermilab, LHC, and a lepton collider with polarized beams such as an NLC or a muon collider; see Chapter 5) to obtain a very accurate answer. The initial data will let us do a crude job, perhaps learning the fraction of the universe to a factor of 2 or so, which is nice but not nearly accurate enough to be satisfactory. The combined data from several colliders will eventually provide the information we need to calculate the answer to an accuracy of 10 to 20 percent, which is good enough to compare with independent cosmology information and to confirm (if it works) that we really understand what the matter of the universe is composed of. If circumstances are favorable, we could achieve that goal within fifteen years.

As is typical of RIP, there is presently a caveat that should be noted. (I include it only so that no one can claim I didn't mention it. And if collider experiments observe superpartners with escaping lightest superpartners, or if the explicit detection experiments observe cold dark matter, this caveat is irrelevant.) It is possible to write a supersymmetric Standard Model in which the lightest superpartner is not a stable particle. Then it could happen that nature is supersymmetric but the cold dark matter is not composed of the lightest superpartner. There are several reasons why I and most others do not think this will happen, but it is a logical possibility. First, the forms of the theory that suggest that the lightest superpartner will not be stable also predict other phenomena that already are known to be false—in particular, they predict that protons should decay in fractions of a second and we should all be gone. If the theory is adjusted to fix this, it often has to be done in an *ad hoc* way without theoretical justification, so the resulting theory may not emerge in a general way. Second, many attempts to make a primary theory, such as some string theories, have properties that lead to the lightest superpartner being stable. Third, even if the general form of the theory does not have a stable lightest superpartner, in some theories the LSP is stable once the world settles into its lowest-energy state. Finally, it would be an awful coincidence if supersymmetry were part of the description of nature but one of its exciting consequences somehow did not occur—in basic physics there are not many coincidences.

The universe seems to contain a lot of forms of matter. We might have thought that everything in the universe is made of quarks and electrons, just as we are, and that the universe also contains photons and neutrinos that move forever, unbound to any star or galaxy. Now we add to that some matter in the form of the lightest superpartner. There are other possible forms of matter that we won't discuss. At our present stage of knowledge, we have to estimate separately how much matter each form contributes to the total, and currently, the amounts of one form seem to be unrelated to the amounts of another. Eventually we expect to do better, and there is already some theoretical effort in that direction: RIP. In principle, the same underlying physics determines the LSP mass, neutrino masses, and the numbers of protons and LSPs and other particles. The earliest models that relate these quantities have been studied in recent years. As we come closer to formulating the primary theory, we expect to obtain a unified understanding of not only which forms of matter make up our universe but also why each inevitably is present in the amount it is.

Higgs Physics

In Chapters 2 and 4, I briefly described the important and somewhat mysterious "Higgs physics" (known by the name of one of its originators, Peter Higgs). In this chapter we will focus on the role played by the Higgs physics in the Standard Model and in supersymmetry. We will examine how physicists hope to discover experimentally the associated Higgs bosons that must be found if the Higgs physics ideas are correct; an experimentally accessible Higgs boson is a necessary consequence of the supersymmetric Standard Model. Appendices A and B provide a further explanation for readers who want more information about how the Higgs physics works and how supersymmetry removes the mystery. The appendices are a little more technical than the description here, but they can be understood without special knowledge.

There are several reasons why the Higgs physics plays a special role in the Standard Model. First, the Higgs boson is a previously unknown *kind* of particle. Electrons and quarks are the matter particles (all fermions). They are all really one kind of particle, differing only in that they carry different amounts of the various charges: electric charge, strong charge, and weak charge. Similarly, the bosons that mediate the Standard Model forces also come in several varieties: photons, gluons, and W and Z bosons. They are all like the photon, differing only in how they interact with the fermions and with each other. All of these particles are the quanta of fields. The fields are all very much like the more familiar electric and magnetic fields whose quanta are the photons. The Higgs bosons, on the other hand, are the quanta of a Higgs field whose origin is not understood—a field that, at least from the point of view of the Standard Model, seems to be different from the other fields.

The Standard Model in its original formulation is a consistent theory only if all of the particles have no mass. If one just adds in masses in an *ad hoc* manner, the theory becomes inconsistent, and it does so in a way that destroys our ability to do calculations and predict or interpret the results of observations. This difficult obstacle was overcome in a brilliant manner by adding to the theory a hypothetical Higgs field that interacts with all the particles that actually have mass, in just such a way as to allow the theory to remain consistent. The resulting theory could describe the masses of the particles and made it possible to calculate observables; the resulting predictions explained much that had been known but not understood up until then and correctly predicted many experimental outcomes. No other way has been found to achieve these goals. A significant aspect of this achievement is that the W and Z bosons, and also the quarks and leptons, must get their mass from the Higgs mechanism, but for technical reasons the bosons and the fermions cannot get mass the same way. The Standard Model including the Higgs physics handles both in simple and natural ways. Technically the Higgs physics works perfectly, and there is little doubt that the Higgs physics provides a correct description of how nature works. But from the conceptual point of view, this aspect of the Standard Model is not understood. Technically the Higgs physics of the Standard Model will probably not change, but how we interpret it, and its implications for extending the Standard Model and strengthening its foundations, are still open issues. One of the exciting achievements of the supersymmetric Standard Model is that it provides an explanation of how the Higgs physics works, and the supersymmetric Standard Model embedded in broader theories that are candidates for the primary theory can fully explain both the existence of Higgs physics and how it works. Some testable predictions that we will examine later in this chapter follow from the supersymmetric explanation.

One facet of why the Higgs physics seems unconvincing even to most physicists arises from the way it handles the W and Z masses. The rules of quantum theory require that a boson that can mediate the forces and is massless can have its spin project in only two directions, whereas a massive boson can point in a third direction. For example, imagine a process that produces a Z boson by colliding an electron and a positron. If the Z were massless, its spin could point only along the electron direction or the positron direction, whereas if it were massive, its spin could point perpendicular to the beam directions as well. Quantum theory tells us to think of each of the states of the Z with different spin projections as different quanta, so the basic formulation of the Standard Model in terms of massless particles has two states of the Z, whereas the massive formulation has three! Where did the other state come from? We can trace it through the mathematics and see that it arose as one of the quanta of the Higgs field— a Higgs boson—that is hiding as the extra spin state of the Z. (When we teach students how this works, we say the Z has "eaten" the Higgs boson and gained weight—that is, mass.) Because all the Z's produced in the real world have this extra spin state, it was hoped that by studying them we could test ideas about Higgs physics. Careful study revealed that was possible, but very difficult. Techniques were devised that made it possible to study the interactions of the extra Z states, and it was shown that aspects of Higgs physics could be tested. But for technical reasons, the features of the interactions that would provide new information were not very apparent, so a large number of the relevant events would be needed, they would have to be events where the interacting Z's were very energetic, and the detectors would have to record the data with great precision. The superconducting supercollider (SSC) whose construction was initiated and then canceled (in 1993) was designed in part to provide the energy and intensity required to perform those tests. Whether other future facilities will ever be able to reach the sensitivities required to study Higgs physics in this manner is unclear.

Finding Higgs Bosons

The most direct way to test the Higgs physics ideas is to produce and detect the Higgs boson explicitly in an experiment. If the supersymmetric interpretation of Higgs physics is correct, that must happen, almost certainly with data that will be taken at Fermilab in the next few years (or possibly even earlier at LEP). Finding a Higgs boson is challenging, not in principle but for practical reasons. First, the very nature of the Higgs boson requires it to interact with every particle with a strength proportional to the mass of the particle; no other particle interacts that way. But the particles we can accelerate to energies large enough to produce a Higgs boson (by converting the collision energy into mass) are electrons, up and down quarks, and gluons (the last three in protons). All of them have very small masses, so the probability of producing a Higgs boson from them is extremely low, millions of times less than the probability of producing W's or Z's. Second, in the Standard Model the mass of the Higgs boson itself is not calculable, so we do not know what mass range to look at, and this makes it a lot harder to find. (In the supersymmetric Standard Model the mass is approximately calculable.) Third, the Higgs boson decays quickly, so only its decay products enter the detector. Higgs bosons are easily mimicked by other kinds of events, so to be sure of a signal, it is necessary to gather a large number of events. In fact, if the Higgs boson has the mass it is expected to have in the supersymmetric Standard Model, dozens have already been produced at Fermilab, but it was impossible to distinguish them from similar events due to normal processes. The present lack of direct evidence for a Higgs boson is understood once one is aware of its special properties and of the practical difficulties involved in producing and detecting it—there is no implication that the Higgs boson does not exist.

Current Evidence

Actually, there is now strong indirect evidence for the existence of the Higgs boson. As we have noted, the Standard Model is an effective theory that can describe many experimental results. The description depends on the masses of the quarks and leptons and bosons, and on their interaction strengths. Once those are all input, dozens of additional observable quantities can be both predicted and measured. Everything we need to interpret all these observable quantities is known, except the mass of the Higgs boson. One can then repeat the calculations a number of times, changing only the Higgs boson mass, and ask which choice of Higgs boson mass gives the best agreement with the measured quantities that depend indirectly on it. The result is that the agreement with experiment is better if there is indeed a Higgs boson, and the best agreement is obtained if its mass is about the same as the Z boson mass. This is not a very precise method; values of the Higgs boson mass up to about two or three times larger still give agreement that is not terrible. But the basic result—that the data imply the existence of a Higgs boson in that mass range—is solid, so there is good reason to expect its direct detection at LEP or Fermilab.

Figure 7.1 shows how an event containing a Higgs boson might look at LEP. It is a digital reconstruction of an actual event in the DELPHI detector. The two isolated lines represent muons from the decay of a Z boson. The two "jets" of particles can both be identified as b-quarks by a detailed analysis, which is consistent with the event's being a collision that produces a Z and a Higgs boson, followed by the Z decaying to muons and the Higgs boson to a b-quark and an anti-b-quark. This is the way a Higgs boson is expected to decay, so the two jets show us how a Higgs boson will look in a detector. Unfortunately, the event is also consistent with the production of two Z bosons, one decaying to muons and the other to b and anti-b. On the basis of one such event, it is not possible to draw conclusions about which of these processes actually happened, but once a number of events are collected, it will be possible to distinguish between the alternatives.

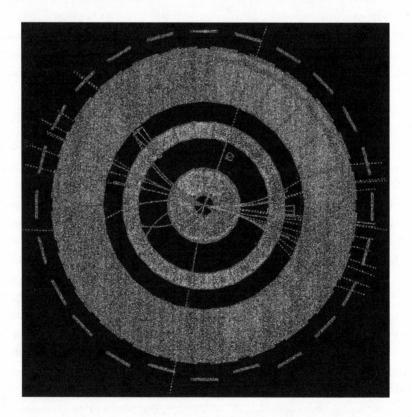

FIGURE 7.1.   An event from the DELPHI detector at LEP, showing how a Higgs boson might appear in the detector. See the text for more details.

LEP, FERMILAB, AND LHC

One of the many ways in which the supersymmetric Standard Model is a better theory than the original form of the Standard Model is that it allows us to make an approximate calculation of the mass of the Higgs boson. In the Standard Model alone, there is no meaningful restriction on the Higgs boson mass. In the simplest form of the supersymmetric Standard Model, we can limit the mass of the Higgs boson to be less than about one and a half times the Z boson mass. That's not a very precise calculation, but the exciting thing is that the entire range can be covered at Fermilab after it takes data with the upgraded collider scheduled to begin operating in 2000. If the collider and detectors operate as expected, this range will be covered in a few years. Data from LEP has already covered the range up to somewhat below the Z boson mass, and it will go a little further before Fermilab accumulates results. It is possible to prove that, even if the real world is described not by the simplest supersymmetric Standard Model but by some more complicated version with extra interactions and particles and unified forces, there is an upper limit on the mass of the Higgs boson of about two Z boson masses, and Fermilab will be able to cover *that* whole range by about 2006. Around then the LHC at CERN will begin to take data that will also cover the whole region, so it may detect the Higgs boson if that turns out to be too difficult for Fermilab, or it can add new information about the properties of the Higgs boson should it be first detected at LEP or Fermilab.

One of the main ways in which we will be able to recognize that a Higgs boson has been produced at a collider will be from the way it decays. Although it can decay into any particle–antiparticle pair, by its very nature as the particle whose interactions generate mass, the Higgs boson must decay into various particles in proportion to their masses. The top quark is so massive that top plus antitop is too heavy to be a decay mode. The b and anti-b are the next heaviest, so they get the largest share of decays. Taus are next heaviest, and the rest are quite light. About 80 percent of the time, a Standard Model Higgs boson will decay into a b quark and an anti-b quark, about 10 percent of the time into a tau and an antitau, and the rest of the time into several other states. The numbers are similar for the supersymmetric Higgs boson (that we actually expect to observe) but in the supersymmetric case they can vary somewhat, depending on several technical details. A decay mode that will be particularly easy to recognize in the detector is muon plus antimuon, and muons are very light; that mode should occur just 1/300 as often as the tau channel. These relative amounts for different modes are very different from those for any other particle, so it will be easy to be sure that a new particle is behaving as a Higgs boson is expected to. For example, the Z and W always have equal amounts of muons and taus in their decays, rather than 300 times as many taus as muons. Even if there are too few events to actually observe the muon decays, it will be easy to check that tau decays occur much more often than muon decays. There are other properties that will provide evidence that it is indeed a Higgs boson that is being observed, but they are less dramatic than these differences in decay modes.

STUDYING HIGGS BOSONS AT FERMILAB

Once the Higgs boson has been detected, presumably at LEP or Fermilab, its mass can be measured. At Fermilab there are two main ways to produce Higgs bosons. One of them produces only about a third as many Higgs bosons as the other, but it produces them in association with a W or a Z that can be recognized in the detector and can therefore serve to mark events that might have a Higgs boson in them. If the mass of the Higgs boson is in its most likely range, the process that produces the larger number will not in practice be useful for discovering a Higgs boson, because other processes happen to give events that can mimic it. But once the mass is known, it will be possible to use all of the events to study the properties of the Higgs boson, recognizing the extra events by their mass. By 2006 Fermilab should have produced well over 50,000 Higgs bosons. That means some very rare decays can be studied. About 20 events of the decay to muon and antimuon are expected, perhaps enough to confirm that the new particle is indeed behaving as a Higgs boson should. Another very important decay is to two photons. This decay does not occur directly but through intermediate states that combine to annihilate into the two photons, and several intermediate states combine to give the result. In the Standard Model there is a unique prediction for this result: about 1.5 events in a thousand—and hence over 75 events at Fermilab, enough to test the prediction. In the supersymmetric Standard Model the result is different, probably somewhat larger, and calculable once the superpartner masses are known.

The bottom line is that if it exists, the lightest Higgs boson of the supersymmetric Standard Model almost certainly should be detected and studied at Fermilab by the year 2006 at the latest (assuming, of course, that the collider is funded to operate as expected, that the collider and detectors function well technically, and that the effort to detect the Higgs boson in the data is given high priority). There are some unlikely but possible ways in which Higgs bosons could behave that would make detection more challenging. For example, if the lightest superpartner was lighter than half the Higgs boson mass, the Higgs boson could decay into two of them. Because the lightest superpartners do not interact in the detector, the Higgs boson would give no visible clue that it was there. But it would still carry off energy and momentum, and knowing this enables us to figure out whether it is actually there.

We have discussed some ways to confirm that a possible Higgs boson signal is indeed the sought-after Higgs boson, particularly by confirming experimentally that it decays into different modes with the expected probabilities, which are different from the way any other kind of particle decays. Once it is detected, several additional properties can be studied. Further, for technical reasons the supersymmetric Standard Model predicts that additional heavier Higgs bosons should exist. They have different masses and differ in some other properties as well. Unlike the lightest Higgs boson, there are no constraints on the mass of the additional ones, so they may be too heavy to be found at Fermilab. Finding and studying them will probably be an achievement of the LHC. If found, they will help demonstrate that the supersymmetric Standard Model is really the theory that applies, and they will provide us with information about the form of the primary theory.

The supersymmetric Standard Model requires that a Higgs boson exist. When (if) one is found, what will it tell us? That depends somewhat on its properties, but probably it will not initially tell us much beyond the fact of its existence. That's because the properties of the Higgs boson are likely to be similar whether it is the Standard Model Higgs boson or the supersymmetric Standard Model Higgs boson. Because the Standard Model Higgs boson is conceptually unsatisfactory, as we noted in Chapter 4, finding the Higgs boson will be taken as a confirmation of supersymmetry, but to be absolutely sure of that, additional evidence will be required. The additional evidence may come earlier, perhaps via the discovery of a superpartner, or later, with the discovery of one of the additional Higgs bosons predicted by supersymmetry. Or perhaps a property of the observed Higgs boson will deviate from the properties allowed for the Standard Model one. If the Higgs boson is found first at Fermilab rather than LEP, for technical reasons it is most likely that the superpartners will have been found first, and once superpartners are found, there will be no doubt that the Higgs boson is the one predicted by supersymmetry.

The role of the Higgs physics in the Standard Model is a necessary one. Some kind of physical effect must occur to play that role. Physicists can be divided into three categories in terms of their attitude about the form the Higgs physics will take. Some—let's call them the fundamentalists—believe there exists a fundamental particle, the Higgs boson, as in the simplest form of the theory. That is what supersymmetry predicts. Others believe there is no fundamental particle at all, but that some as yet unknown form of the interactions among particles at higher energies will somehow play the role of the Higgs physics—they are the atheists. Some atheists have invented very sophisticated theories, invoking unknown quarks and leptons interacting with unknown forces in just such a way that some of them bind into bosons that are just like Higgs bosons, all to avoid having Higgs bosons that are as fundamental as photons and electrons. The third group consists of agnostics—those who are uncertain. The arguments among members of these groups have been fruitful, because they have led physicists to think of and work out better experimental ways to learn how nature actually solved the Higgs physics problem. If the answer is found in data at Fermilab by about 2005, it will have taken about four decades since the problem was originally formulated—typical of the amount of time it usually takes to answer basic questions in physics. When we know the form the answer takes, it will not only complete the Standard Model but also help us discover how to extend the Standard Model. If the supersymmetric Standard Model is indeed the next stage toward the primary theory, there is a Higgs boson that can be detected at Fermilab.

8

Some Additional Help from Supersymmetry, and Some Challenges

If nature is really supersymmetric, supersymmetry should affect additional physics issues, even beyond the main ones described in Chapter 4. There are a number of phenomena we hope to understand that cannot be explained in the Standard Model, such as the matter–antimatter asymmetry of the universe (mentioned in Chapter 2 and described more fully below—we'll abbreviate the name to *matter asymmetry*). If supersymmetry is indeed part of the underlying laws of nature, it could aid—or even be crucial for—our understanding of the matter asymmetry and other phenomena. In this chapter, we'll consider a few topics where supersymmetry does seem to offer the possibility of new explanations, and I'll mention some phenomena for which supersymmetry seems to do less well than the Standard Model.

In Chapter 4 we looked at an impressive list of basic issues where supersymmetry might play a crucial role: explaining the Higgs physics, solving the problem that the Planck scale is very different from the scale of the weak and strong forces (the "hierarchy" problem), allowing a unification of the forces, and accounting for cold dark matter. Some extension of the

Standard Model was essential, yet whatever that new physics might be, it did not affect the precise tests of the Standard Model carried out at LEP and other colliders. Supersymmetry was not introduced or designed to solve any of these problems, but once the idea of supersymmetry arose, people studied the theory and realized that it had these implications. In this chapter I'll describe a few more phenomena where supersymmetry may play a major role. These are all very much RIP. The supersymmetric aspects of all the explanations are tightly linked to each other, so any additional effects of supersymmetry are strongly constrained by the other successes—if additional phenomena are explained, it is significant. We do not yet know the actual explanations for the phenomena described in this chapter, but the new approaches based on supersymmetry are leading to new ideas in these areas and have testable implications.

Matter and Antimatter Asymmetry

Can we ever understand how our universe came into existence and why it is the way it is? If we can, one of the properties of the universe we will have to explain is the matter asymmetry. The crucial point is the "asymmetry." We might think that we could hope to understand the origin of the universe if all of its properties had a net value of zero. If the universe had a net electric charge, for example, or a net total energy, where did those net values come from? But if the net values were zero, then perhaps an explanation would be possible. If the universe emerged somehow from nothing, it would have to have no net amount of any quantity. For electric charge, there is good evidence that there are as many positive charges as negative ones and that the net is indeed zero. That's true for energy too, when it is properly understood. Energy can have positive or negative values. If you have water above a water wheel, you can let the water fall on the wheel and turn it and do work for you. But if the water starts out below the wheel, you have to do work to raise the level of the water up to where it can again do work for you. The energy is positive in one case, negative in the other. It's positive if you can get work out of it, negative if you have to do the work. Most of the negative energy in the universe is in the form of objects such as stars bound by mutual gravitational attraction to one another in galaxies—you would have to do work to tear a star away from a galaxy. A lot of the positive energy in the universe is in the kinetic energy of the moving stars, and some is in the form of the mass of stars and other objects. If you calculate how much there is of each and add them, the total energy comes out to be zero.

Thus, when we study different properties of the universe, we find that the net values seem to be zero and therefore consistent with the idea that we can hope to explain the origin of the universe. But there is one property that is clearly not zero: the net number of neutrons plus protons, the matter of the universe. Because the net electric charge of the universe is zero, there is an electron for each proton, so we'll ignore the electrons in the discussion; whatever explains the proton excess will explain the electron excess. When energy is converted into mass, it always makes equal numbers of particles and antiparticles. If the universe began from nothing, we would expect it to have equal amounts of matter and antimatter. Where did the excess come from?

First let's ask how we know that there is an excess of matter. If some galaxies were matter and some antimatter, when such galaxies occasionally collided, the particles would annihilate with one another and produce photons of several predictable energies that we could detect. Although efforts have been made to find such photons, none have been seen. More generally, photons from matter–antimatter annihilation would be relics we could detect, and they would typically have properties (such as their energies) different from those of photons left over from the Big Bang. We do see the photons from the Big Bang, and they have the expected properties, without any modification due to photons from matter–antimatter annihilation. These arguments imply that at least most of the universe, and probably all of it, is matter, not antimatter.

For many years, this seemed to imply that we would not be able to understand how the universe began. Then the great Russian theoretical physicist and political dissident Andrei Sakharov pointed out in 1967 that if certain conditions were satisfied, it would be possible to have a universe that began with equal amounts of matter and antimatter but evolved to have matter dominate. Basically there were three conditions that would all have to be satisfied. For our purposes, they can be stated as follows. First, it was essential that protons and antiprotons not be stable, permanent things that would live forever. Second, there had to be some interaction that treated matter and antimatter differently. Third, the universe had to be cooling and expanding as time passed.

The third condition had been known to be satisfied since the discovery of the expanding universe in the 1920s. To understand this condition, suppose that some process did occur that led to a matter asymmetry. For example, perhaps some as yet unknown particle and its antiparticle were produced during the Big Bang, and they decayed in a way that produced more matter than antimatter. If the universe were static, eventually the decay products would collide and produce the original particles, so there would be an equilibrium, and no net matter asymmetry would be produced. But if the universe expanded and cooled before the decay products collided, the products would have less energy and would not make as many of the original particles, so a net matter asymmetry would remain.

In 1954 it was found that the second condition held for one experimental reaction, so it was hoped that it would hold more generally. In 1967, the year Sakharov enumerated the conditions for matter asymmetry, there was no known example of a process meeting the first condition—indeed, there was considerable doubt that such a condition could be incorporated into a consistent theory. Then in the 1970s, people began to write theories that unified the description of quarks and leptons, and after a while they realized that if quarks and leptons could be understood as different facets of the same object, then they could turn into each other in certain ways. This would imply that protons and antiprotons would decay—and the first condition would be satisfied. At that stage it was clear that there was finally an opportunity to solve the matter asymmetry problem.

Two decades of study have shown that there are actually several ways in which the matter asymmetry could appear as the universe evolved. At the present time, we do not know which is the actual way; it could even be that more than one contributed. Basically, today we know two things: (1) It is possible in the context of current theories to explain the matter asymmetry. (2) This cannot be done in the Standard Model alone but, rather, requires that the Standard Model be extended. There are presently three main ways by which the observed matter asymmetry may have come about. Supersymmetry plays a crucial role in two of them and an indirect role in the third. The issue of the matter asymmetry is RIP, so we do not yet know the final explanation, but there is good reason to believe we will after more analysis and after testing all the possible methods, each of which generates other associated predictions. For example, one proposal is that the matter asymmetry was generated rather late in the evolution of the universe. It can give a satisfactory numerical value for the matter asymmetry only if the Higgs boson exists with a mass small enough to be easily observed at Fermilab, and if, in addition, the stop (superpartner of the top quark) has a mass less than that of the top quark, in which case the stop is also easily detectable at Fermilab.

PROTON DECAY?

Let us turn to the question of whether protons are stable or can decay. One of the first people to think about it was Maurice Goldhaber. He pointed out that the simple fact that we don't all die of cancer at an early age enables us to deduce that the proton has a lifetime much longer than the age of the universe. If a proton in a living body did decay, it would release enough energy to damage thousands of molecules throughout that body. That would not be so bad if it were rare, but if it were common (there are about 10^{29} protons in a body), it would cause cancer. We would die by our teens or twenties, which leads to Goldhaber's conclusion.

The Standard Model implies that protons do not decay at all, but it does so in a way that on examination seems to be accidental rather than a consequence of a general principle. And that turns out to be the situation—in extensions of the Standard Model, proton decay is possible. The first estimates, in the middle 1970s, predicted a proton lifetime longer than Goldhaber's limit and led to some impressive experiments in search of the decay. If protons are going to live much longer than the lifetime of the universe, you could get really bored watching one to see if it decays. But in a world described by quantum theory, the timing of decays is random. Any decaying object, such as a radioactive nucleus or a proton, has a "half-life." This means that half of any sample will decay during the half-life of that substance. From the time you start watching until the half-life, they will decay at a uniform rate. Thus, if enough protons are brought together and watched, some of them will be observed to decay if the proton is not stable. If the half-life is very long, the sample has to be very large in order for the observer to see one decay in (say) the first year of watching.

The first large experiment was done by constructing a tank of water the size of an Olympic swimming pool. The entire inside of the pool had to be surrounded by phototubes sensitive enough to pick up any photons (light) emitted if a proton decayed. The water had to be so pure that the light from a decay would not be absorbed as it crossed meters of the water. Because cosmic rays can mimic a signal, the pool had to be deep underground—in this case, in a salt mine near Cleveland. That experiment did not find evidence for the decay of protons, and today an even larger tank is being watched in Japan.

The argument that protons might decay comes from unified theories that suggest that at sufficiently small distances, there are additional forces that can turn quarks into leptons, in a manner analogous to the way electrons and neutrinos can turn into each other (as illustrated, for example, by Figure 2.2), but with much less probability. For example, in some unified theories, two quarks can turn into an antiquark and a positron. The proton is a tightly bound state of three quarks, so if two quarks turn into an antiquark and a positron, the antiquark will quickly annihilate with the third quark, giving most probably two photons (plus the positron) that can travel through the water. The positron will quickly annihilate on an electron, also giving two photons in the annihilation, so the net result will be four photons with energies adding up to a proton's mass. The earliest ideas predicted a lifetime that the experiment in Cleveland could detect. But even before that experiment reached its peak performance, some people realized that the unified theory should really be a supersymmetric one for full consistency and that, for technical reasons, the supersymmetric unified theory predicted a lifetime about a thousand times longer than the original prediction—probably too long to be observed in the initial experiment. The experiment in Japan is just now beginning to test the supersymmetric predictions. Perhaps a signal will be reported early in the new millennium.

The issue of whether protons decay with a lifetime short enough to be observed eventually is a case where the prediction of the theory is at present simply unclear. Some versions of the theory do have protons decaying, and some do not. If protons are seen to decay, there are a number of additional things we can learn about how they decay that will teach us about the basic theory. For example, different forms of the theory and different ways of breaking supersymmetry predict different proton half-lives. The decay could occur in ways different from the one described above, and the relative amounts of the various ways would tell us a lot: One mode is the positron and two photons mentioned above, another is a muon and two neutrinos, a third two positrons and an electron, and so on. Simple supersymmetric unified models predict that the muon-and-two-neutrino mode will occur most frequently.

The experiment in Japan probes part of the lifetime range of the supersymmetric unified theory, but not all. Investigators planning another experiment in Italy hope to probe more of the range. If they do not find the proton decay, it may be that we will never know experimentally whether the proton decays. The necessary experiments may be too large and expensive ever to be done. They are limited mainly by the need to get such a large number of protons together and to be able to watch each one.

RARE DECAYS

Another area where supersymmetry changes how we view the issues is "rare decays." In the Standard Model, there are a number of decays that are predicted not to occur at all. An example is $\mu \rightarrow e\,\gamma$. Because the masses of the electron and the photon are smaller than the muon mass, energy conservation allows this decay to occur. But there is no way that the vertices of the Standard Model (Chapter 2) could be combined to lead to this decay, so the Standard Model predicts that it does not occur. So far, more than a thousand billion muons have been examined in an effort to find any muon decaying into electron plus photon, and none do. But when the form of the theory is examined to understand better the basis for this prediction, it seems to be rather accidental, without a fundamental basis. That it is accidental becomes more likely when the question is examined in the supersymmetric Standard Model, where the decay *is* predicted to occur. If one combines the allowed vertices in the supersymmetric Standard Model, it is possible to draw a somewhat complicated Feynman diagram that begins with a muon and ends with an electron and a photon. Once there is a Feynman diagram, there is always an associated probability for the process to occur. Finding this decay and analogous ones would constitute additional indirect evidence for supersymmetry.

Indeed, even before there was any experimental check, one could have made an estimate in the supersymmetric Standard Model of how often this decay should occur. Then the naïve prediction would have been that such decays were expected to occur in the experimental sample that now exists, but none were observed. We don't yet understand how to interpret this. Of course, it could be a clue that nature is not supersymmetric at all, and that, consequently, reasoning based on supersymmetry isn't meaningful, but we'll keep that possibility stored at the edge of consciousness while not entirely discarding it. The expectation that decays should have occurred surely implies that improved experiments that observe more muon decays should finally detect the e + γ decay. On the other hand, the absence of this decay has implications for how supersymmetry is broken, and investigators often try to use it as a test of ideas about how supersymmetry is broken.

This is not the only such rare process. Other processes involving taus and quarks also occur less often than expected. If they are ever observed, it will be a qualitative triumph for supersymmetry, though because there are not superpartners in the final state, the role of supersymmetry is an indirect one and other interpretations will be possible. Data on how often such decays occur will provide important information on how the supersymmetry is broken and about the form the theory actually takes.

The experiments to search for such rare decays are heroic. The experimenters have to examine more than a thousand billion decays to find a few of the ones that will tell us something new. To do that in a year, they have to look at over a million decays each second, never miss one of the interesting ones, and never be fooled by a typical decay where the particles happen to be in an improbable but allowed configuration, such as two overlapping so that they look like one, or one moving so slowly that it is invisible to the detector. It is essential to understand the detector's behavior far better than we usually understand anything in our environment.

CP VIOLATION

Another area I'll describe is an extremely fascinating one called *CP violation.* C stands for the operation of replacing every particle with its antiparticle (and conversely) in a process, and P stands for looking at a process as in a mirror; the jargon is that C is from "charge conjugation" and P from "parity inversion," but we won't use those words. In the Standard Model theory, if you take any process that can occur, replace all of the particles with their antiparticles and conversely, and reflect the process as in a mirror, you get another process that can occur. If the two processes occurred with equal probability, we would say that CP is conserved. For one set of particles (kaons), the measured probability of the two processes is almost but not exactly identical, differing by a few parts in a thousand, so we say that CP is violated. CP violation is also one of the conditions necessary to obtain a matter asymmetry, as described earlier in this chapter, so the existence of the matter asymmetry provides a second place where CP violation has occurred, an indirect observation. So far these are the only two instances. We also know that the strength of the effect in the kaons is not large enough to account for the observed matter asymmetry, so apparently these are two different CP violation effects, unless both arise from one underlying origin. One of the main purposes of the new facilities called b-factories, which started taking data in 1999 at the Stanford Linear Accelerator Center, at the KEK Laboratory in Japan, at Tsukuba near Tokyo, and also at the upgraded facility CESR at Cornell, is to look for CP violation effects in the b-quark interactions. In addition, the CDF detector at Fermilab can identify b-quarks and is searching for CP violation effects as one of the many areas it can study by selecting different subsets of the data it takes, and there is an experiment under way at the German laboratory called DESY in Hamburg. If we can learn the form the CP violation effects take for b-quarks, it should greatly help us to untangle their origin.

Supersymmetry allows a number of interactions involving superpartners to show CP violation effects. If and when superpartners are produced, it will be very interesting to look for these interactions. More important, supersymmetry is capable of producing the CP violation that is necessary to explain both the matter asymmetry and the small amount of CP violation observed in the Standard Model, whereas the Standard Model is not. It would be exciting and esthetically attractive if all of the CP violation effects originated in the supersymmetric interactions rather than in several unrelated mechanisms. That is a plausible and testable possibility.

INFLATION

The last area to remark on here is *inflation.* There are strong arguments that the universe went through a brief period of extremely rapid expansion, called inflation, very early in its history, just before the Big Bang. One can think of the universe as being mainly full of energy during the inflation—energy that exerted a large expansionary pressure. At a certain stage, this energy was radiated away into a huge number of particles of all kinds. That eruption into particles was the Big Bang.

The energy density was in the form of a field, or several fields, dubbed inflatons. Inflation has been studied largely in terms of generic fields that might do what is needed to make inflation happen, with little consideration of whether such fields really existed or of their physical origin. Supersymmetry can supply such fields. Many of the spinless fields of supersymmetry have properties consistent with what is needed to provide real physical fields for the inflatons, rather than treating them as hypothetical fields with no independent reason to exist. Work on this subject is very recent, and we do not know whether it will turn out to be how inflation really works, but it is another area wherein supersymmetry can play a role, even though that's not why supersymmetry was invented.

PERSPECTIVES
AND CONCERNS

For all of the phenomena that I have described so far, supersymmetry extends the Standard Model and improves on the Standard Model in many ways. There are, however, three aspects for which the supersymmetric Standard Model apparently does *not* do as well as the Standard Model without any extensions. Basically, they are phenomena that do not happen in the real world, and that are expected not to happen according to the Standard Model theory, but that the supersymmetric Standard Model predicts could happen. One is rapid proton decay, occurring in a fraction of a second rather than billions and billions of years. Another is certain decays or transitions of muons and taus and quarks. The third is too-large CP violation effects. (For those who like to know the jargon, these are baryon and lepton number conservation, the flavor changing neutral current problem, and the supersymmetry CP problem.) The details are rather technical and we can skip them. The important thing is that there is no way in the Standard Model for these effects to occur, but in the more general supersymmetric Standard Model, they are possible. We may have to put apparently *ad hoc* restrictions on the supersymmetric Standard Model to make it consistent with the absence of these effects, though this is not certain. Are these clues that something is wrong with the supersymmetric Standard Model in spite of its many successes?

There is at present no consensus about the answer to that question. My view has two parts. First, I do not think that the success of the Standard Model in avoiding these effects is significant, because it seems to be accidental. That is, no principle ensures that the Standard Model will not allow these effects. But the Standard Model is an effective theory (recall Chapter 3), not the fundamental theory, and we can trust as basic aspects of the Standard Model (or of any theory) only aspects that are there because of general principles. Second, there are several ways in which general principles could make the supersymmetric Standard Model consistent with these three effects. If we think of the supersymmetric Standard Model also as an effective theory whose form is determined by viewing the primary theory in our four-dimensional world at distances larger than the Planck scale, it is easy to imagine properties of the primary theory that would constrain the supersymmetric Standard Model to get these effects right. That is, if there were no way to add principles to the supersymmetric Standard Model that would make it behave appropriately here, then one should be very concerned, but because there *are* such ways, these apparent weaknesses are probably really clues to the structure of the primary theory and to how supersymmetry is broken. Thus I (and many others) actually view these "problems" as opportunities for insights into the primary theory and its connection to supersymmetry in our world.

Because supersymmetry is an effective theory rather than the primary theory, there must be issues it cannot explain, and there are. Let's conclude this chapter by mentioning some of them. The next chapter describes the relationships among supersymmetry, string theory, and the primary theory, where these issues can be addressed. Supersymmetry does not tell us anything about why the quarks and leptons come in three families or about the numerical values of the masses of the quarks and leptons. We saw earlier that with supersymmetry the Standard Model forces can unify, and we also saw that supersymmetry requires some connection with gravity, but supersymmetry does not explain *why* the forces should unify or why they should unify with gravity. In addition, although the ratios of the strengths of the forces are determined if they unify, the overall magnitude of their strengths is not fixed by supersymmetry.

One of the most profound problems in physics is called the *cosmological constant problem*. Whenever one writes the equations that govern an area of physics, one writes the most general equations that are allowed by the principles and the rules of quantum theory and relativistic invariance. In the case of gravity, there is a certain term allowed in the general equations, called the cosmological constant (it was mentioned briefly in Chapter 1). If that term is in the equations that describe gravity, it can be interpreted as giving the universe an energy density that is constant in space and time. That energy density pushes out on the spacetime of the universe and accelerates the expansion rate of the universe—the larger the cosmological constant, the faster the universe should expand. One can estimate the maximum possible value of the cosmological constant from the observed expansion rate of the universe. We can also estimate naïvely what size the cosmological constant should be if we made the simplest guess. The problem is that the maximum size the cosmological constant could have, and still be consistent with what we observe, is many powers of 10 smaller than the naïve estimate.

Something is wrong with our understanding. If we were wrong by a factor of 2, or even a bit more, we might expect that when we understood the theory better and could calculate instead of estimate, it would work out all right. But when we are off by orders of magnitude, there must be a missing principle—and a very basic principle that affects our understanding of the structure of the universe. Supersymmetry actually can help in a sense, because if the supersymmetry is unbroken (as it is in string theory—see Chapter 9), it is possible for the cosmological constant to come out to zero. In the supergravity theory with broken supersymmetry, the cosmological constant can still be zero, but it has to be fixed at zero instead of automatically coming out to be zero, which is not satisfactory. If it is not set to zero, the experimental value of the cosmological constant and the naïve estimate differ less than before, but still by far too much. Many string theorists believe that once we understand how supersymmetry is broken, we will also understand the cosmological constant problem, and that may well be so, but it is not guaranteed.

Recently this problem has been refined by better data on the expansion rate of the universe. Apparently the expansion of our universe is accelerating a little. If so, that means the cosmological constant is not zero, though it is still far smaller than the naïve estimates. That may make it more difficult to solve the problem—one can imagine a principle that gives zero for the cosmological constant. It also is puzzling because the amount of energy density that the cosmological constant supplies to the universe is very close to the energy density supplied by all the forms of matter, and we don't know why this should be. This is an area where planned satellite and balloon experiments (called MAP and PLANCK and others) will provide, over the next decade, data that will greatly help us understand how the energy density of the universe is distributed.

Finally, there is the crucial issue of why supersymmetry is a hidden, or broken, partial symmetry. Just as the Standard Model could not account for how and why *its* symmetries were broken ones, so supersymmetry cannot provide an explanation for its own breaking. That explanation must come from outside of supersymmetry.

.... **9**

Supersymmetry, String Theory, and the Primary Theory

This chapter could have two subtitles: "Why I prefer to be a supersymmetry theorist rather than a string theorist" and "String theory is indeed testable, and if the world is supersymmetric at the collider scale, then string theory is testable even in the conventional historical sense." Let's start by recalling the goals of particle physics. What we want is first to learn what the laws of nature are and whether they can be unified into one or a very few basic principles. We want to know where the laws of physics came from and why they are what they are. We want to find the primary theory (recall the discussion of effective theories in Chapter 3). We want to explain our world in particular, why there is a universe, and why it is the way it is. In order to do all of that, we must not only understand the particles, the forces, and the rules, but also learn what spacetime is and why nature's rules for calculating the behavior of particles for all forces are quantum theory and relativistic invariance. We have to understand why the quanta of energy have a minimum size (why Planck's constant, h, is not zero) and why the speed of light, c, has an upper limit; what is important is the fact that h is not zero and c is not infinite, not the actual values of h and c. And we have to understand whether the primary theory is unique—the only possible theory of its kind that could be mathematically consistent.

As is explained in Chapters 2 and 3, we have basically succeeded in understanding how our world works at all scales down to near the Planck scale (assuming that supersymmetry is indeed discovered). We know the basic particles, the quarks and leptons; and we know the forces and how they operate, mediated by the exchange of the bosons γ, W, Z, g, and G. We do not yet know why the world works this way. It's important to be aware of this distinction between "how understanding" and "why understanding." In physics the "how" came first, and now we want to have the "why." That order is inevitable, given that we are after the primary theory, but it can be different in other areas of science. For example, in biology the fundamental organizing principles (the "why") are two: evolution by natural selection, and the genetic code based on chemical molecules. All of the diversity and properties of life on earth can be traced back to the first self-replicating molecules, having evolved via the differential survival of organisms in a changing environment. All the changes in organisms can be traced to changes in their genetic structure. However (for example), we do not yet understand in molecular detail how an egg plus a sperm develops into a person with eyes and fingers and a brain, though much has been learned about development in recent years. Thus biology has the "why"—the basic framework—but how the principles actually lead to the phenomena that we observe is still being worked out.

String Theory and M-theory

Even if it is correct, string theory is not the primary theory. To understand this, we must first discuss what string theory is. String theory incorporates quantum theory, relativistic invariance, gravity, and the Standard Model forces and particles into one consistent theory. To achieve that goal, it has been necessary to formulate the theory in ten spacetime dimensions (rather than the three space plus one time dimensions of our everyday world) and to interpret the particles as string-like objects rather than point-like objects as in the Standard Model. But string theory is much more. Basically, in the ten-dimensional world, there is only one universal force, gravity. When three of the space dimensions are large compared to the other dimensions, the string theory world should look like our world, with our forces. We don't yet understand how or why this happens, but string theorists have shown that it can happen. Further, there are several fundamental questions that we know cannot be fully answered in the supersymmetric Standard Model, such as why there are three families of particles, and why the particles have the masses they do. String theory provides answers to these questions, but we do not yet know whether it provides the right answers, because so far it has not been possible to work out the answers quantitatively. To have a theory that addresses those questions at all represents major progress! The essential reason why string theory should be studied as thoroughly as possible and taken extremely seriously is that it is the first theory in history that can fully include, or perhaps imply, gravity and quantum theory and the Standard Model forces and special relativity and can address questions such as why there are three families.

However, string theory does not address the even deeper issues of defining spacetime and showing why the rules must include quantum theory. Rather, it assumes a background of space and time, and it assumes the rules. Some aspects of string theory suggest that it is close to being able to address these issues, but an extension will be needed. Also, string theory is not unique—there seem to be several versions that apparently can predict different phenomena. This latter problem has been dealt with in the past few years, leading to the next stage, called M-theory. (There is not yet agreement about what "M" stands for—suggestions include *magic*, *mystery*, and *mother*.) M-theory is formulated in eleven spacetime dimensions, and it has been possible to argue that the five apparent forms of string theory in ten dimensions were in fact different ways of looking at M-theory in eleven dimensions. That insight may achieve the goal of having a unique theory. All of this is very recent, and very complicated, so it will take time to sort out just what is achieved. It may be that with these developments, the search is converging on the primary theory, or we may still be far from that stage. M-theory usually is viewed as assuming spacetime and quantum theory (rather than deriving their existence), in which case it is not the primary theory either.

BROKEN OR HIDDEN OR PARTIAL SUPERSYMMETRY

In order for string theory to be consistent, it is probably necessary for it to be supersymmetric. It must have full supersymmetry, not a broken supersymmetry (Chapter 4). This is a very important distinction. We expect two things: the ten-dimensional (nine space plus time) string theory near the Planck scale is fully supersymmetric, and the four-dimensional (three space plus time) supersymmetric Standard Model at the collider scale has the supersymmetry hidden or broken. When the supersymmetry is hidden, it is not totally invisible—the superpartners still exist, but the particles and their superpartners can have different masses and some modified interactions. Different ways of breaking the supersymmetry lead to different patterns of masses and interactions, so when the masses and interactions are measured, we will have experimental information pointing to how the supersymmetry is broken.

Why or how the supersymmetry is broken is simply not yet known. It may be helpful to recall that it was also necessary to break the particle interchange symmetry of the Standard Model before the Standard Model could correctly incorporate mass into its description. Figuring out how that worked was difficult and resulted in introducing the Higgs physics. An analogous effort is likely to be needed to account for the supersymmetry breaking.

THE ROLE OF DATA

String theorists work at the Planck scale, and their theories have unbroken supersymmetry, though they keep an eye open for how the supersymmetry might be broken. There are ideas that may explain how the full supersymmetry at the Planck scale could become a broken or partial supersymmetry at larger distances. Part of the problem is that we don't yet have data on the superpartners, so even if one of the already discovered methods to break supersymmetry or curl up extra dimensions is right, we have no way of knowing. Supersymmetry theorists like myself also think about physics at the collider scale, working out how to recognize that superpartners have been produced and how to take the quantities experimenters actually measure (such as the probabilities for various processes to occur and the directions and energies of the particles that emerge into the detectors), and convert that information into the patterns that can be compared with possible implications of string theory. We do this because we think it is very likely that simply thinking about the world at the string theory level (or the M-theory level, or the primary theory level) will not lead to a convincing way to go from the ten-dimensional supersymmetric theory to the four-dimensional broken-supersymmetry theory. We expect it will first be necessary to use insights from data—that is the way physics has always progressed historically. And that is why I would rather be a supersymmetry theorist.

The reason why the (expected) discovery of experimental evidence for (broken) supersymmetry and the data about the properties of superpartners will be superimportant is that these investigations open the window onto a view of physics at tiny distances like the Planck scale. Making the connection between the ten-dimensional unbroken-supersymmetric string theory near the Planck scale and the four-dimensional broken-supersymmetric Standard Model at the collider scale probably requires addressing three major issues. One is finding out why three space dimensions are large and the rest are small (this is usually spoken of as the "compactification" problem, even though what needs explaining is that some dimensions are large) and how the small dimensions are curled up into spheres, or donuts, or some other shape (their "topology"). The second issue is finding out how the supersymmetry becomes a broken, partial, hidden symmetry—the symmetry is still there in that the superpartners exist and all the required interactions exist, but the numerical values of the masses and some related quantities are different from what they would be in the unbroken-supersymmetric theory. The third issue is finding the state of lowest energy of the universe, the state it settles into. The theory currently seems to allow the universe to settle into many possible states, and they seem to lead to different predictions. These three questions may be connected (for example, the same mechanism that causes some dimensions to become large could break the supersymmetry), or they may be independent. As in the past, once we learn the actual description of nature, understanding will soon follow.

EFFECTIVE THEORIES AND THE PRIMARY THEORY

The data to guide thinking are data from three space dimensions, reflecting (presumably) partial supersymmetry, whereas the basic theory has ten space dimensions and complete supersymmetry. Remember the discussion of Chapter 3, where we learned that each level of understanding of nature can be described by an effective theory. It may help to visualize the progression of effective theories as stages on the way to the primary theory, as in Table 9.1. Since M-theory is not yet fully defined, I am separating it from the primary theory because M-theory usually treats spacetime and quantum theory as givens. It could turn out, however, that M-theory and the primary theory merge. Of course, until we know the answers, any such table has to be tentative.

TABLE 9.1.   From Effective Theories to the Primary Theory.

The Standard Model

The supersymmetric Standard Model
(valid from the collider scale to the scale where forces
are unified—hidden supersymmetry)

String Theory
(several ten dimensional limits of M-theory—full supersymmetry)

M-Theory
(eleven dimensions, full supersymmetry, possibly unique, assumes quantum
theory and space-time (?))

The primary theory
(includes a derivation of space-time, and the meaning and number of dimensions, explains why quantum theory and relativistic invariance are the rules
of nature, where the laws of physics come from, and why M-theory is the
unique theory describing our world)

The higher layers of effective theories at larger distances than the Standard Model were described in Chapter 3. These effective theories are at different stages of completeness. The Standard Model is complete except for
the Higgs physics (Chapter 7). How the Standard Model is extended to include neutrino masses, and where this fits in the table, are presently uncertain. The supersymmetric Standard Model, string theory, and M-theory
are very much RIP; a great deal has been accomplished, but there is still a
long way to go before they are established or fully worked out, although
progress could be rapid if there is data on superpartners. Addressing the
questions of the primary theory has begun.

Can We Really Understand the Origin of the Universe and Its Natural Law(s)?

Suppose that the progress optimistically anticipated in this book actually occurs. Superpartners and Higgs bosons are indeed detected and studied experimentally. Their properties point the way to a string theory whose four-dimensional form explains all the features of the world of particles and cosmology. How much further can we hope to go in understanding where the laws of nature and the universe came from? Are there limits? Even if we can eventually understand the universe, is it premature to hope to do so soon? People have argued that because the primary theory should be about nature at the very tiny distances of the Planck scale, which we can never directly probe, we can never test ideas there—and thus string theory may be a lovely idea but is not science. Could that be so?

As always with science, we can't know what we can understand until we get there, so we will be certain of the answers to these questions only if we succeed. But we can look at the arguments against the possibility of learning and testing the primary theory and see how good they seem to be. The outcome is that the skepticism is not backed up by sound arguments.

First, let's consider whether we can hope to discover the primary theory. Remember, the goal is to *understand our* world. Both italicized words are crucial. People have raised a number of issues meant to suggest that we will not achieve that understanding. One argument is that there is much that we cannot know about the world and that science has added to that list. For example, because of the finite speed of light and (consequently) of information, we cannot know what is happening right now on Mars, let alone halfway across the universe. We cannot know the position and momentum of a particle simultaneously to better than some accuracy set by the uncertainty principle. But our goal is not to *know everything* about the universe but rather to understand how the universe works and why it is the way it is. Learning that there are things we cannot know (as in the foregoing examples) is actually part of that understanding, and in no way does it imply that we cannot understand how the universe works and why it is the way it is. We cannot know all possible chemical molecules, for example, but we can fully understand the principles that govern the formation and behavior of all possible molecules.

Another argument is based on Gödel's incompleteness theorem. This astonishing and elegant mathematical result was proved by Kurt Gödel in the 1930s. It basically states that in any mathematical theory that is interesting for our purposes, there are true results that cannot be mathematically proved to be true, and also that you can't prove the consistency of a mathematical system from within that system. For mathematicians, that has profound implications. A number of people have worried that it also means that physicists would be unable to show that the primary theory applied to the world even if it did. But that is not how science works. There are two important differences.

First, we do not need to prove all possible theorems, nor do we have to prove the consistency of the whole system. We already have our world, and we know it is described by consistent laws—otherwise, it would fall apart. If the equations for the stability of atoms changed with time or were inconsistent with the equations for the forces, atoms would not keep existing and forming the world. Only consistent equations have solutions. (Indeed, it is often remarked that it is "amazing" that our world is comprehensible scientifically. But is that really surprising? Our world must behave according to mathematical regularities if it is to exist for some time. Given that these regularities exist, we can learn what they are.)

Second, scientific results are never proved to be true. "Proof" is for mathematical theorems. At a certain stage of research, the evidence for a given result becomes so strong that it is accepted by those who understand it.

Every result depends on certain parameters (such as distance or speed). If the evidence comes only from a limited range of those parameters, the result may or may not change when it is extrapolated outside that range; recall the discussion of Chapter 3. For example, the laws of gravity have not been tested for distances smaller than about a millimeter. Experiments are under way to determine whether the form of the gravitational force changes at smaller distances, as motivated by some ideas from string theory and extra dimensions. On the other hand, our descriptions of all forces have now been tested for all speeds, from rest to the maximum possible speed (that of light), so there will not be further modification there. In addition, as we have seen, accepted scientific results form a coherent structure with many implications that strengthen our confidence in them. The results of science lead to our surest knowledge about our world because of the process used to obtain them, and this process includes improved experimentation and consensus among informed workers. The bottom line is that Gödel's theorem is simply not relevant to whether we can understand how our particular world works and why it is the way it is.

Yet another concern arises from a feeling of humility in the face of the awesome size, complexity, and beauty of the universe from the particles to the cosmos. How can mere humans expect or even hope to understand all of that and how it originated and why it is the way it is? It is sometimes said that it would be like expecting a dog to understand quantum theory. Darwin first used that analogy. He wrote, "I feel most deeply that the whole subject is too profound for the human intellect. A dog might as well speculate on the mind of Newton." It is easy to understand why Darwin would feel that way, living as he did before the powerful discoveries of the past century. We won't know whether humans can figure out the primary theory unless our efforts to understand either succeed or hit a dead end. The approach of science since it began has simply been to try to understand natural phenomena as well as possible and to see how far we can go. Personally, I think that if a dog were able to get data about atoms and inquire about how they work, then that dog could discover quantum theory, so I remain hopeful that we will indeed come to understand the universe.

TESTING STRING THEORY AND THE PRIMARY THEORY

Could we fail to understand the world because we cannot test ideas? This is a more subtle and interesting point. Some people have suggested that because we can never build a collider that can directly probe the Planck scale, we can never test ideas about physics there or test the primary theory. That is simply wrong. One of the ways in which people who argue that there are not tests are confused is in thinking that it is necessary to have been somewhere to know what happened there. We do not have to have watched the Big Bang to establish that it occurred, because several of its consequences affect our world and are fully testable. We do not have to have been there when the dinosaurs became extinct to figure out how that happened—again, there are clues that enable us to unravel the mystery. We do not have to travel near the speed of light to figure out and confirm that it is the maximum speed at which we can travel. There are always clues and relics that will help us test ideas. Finding those clues and relics requires dedicated effort by talented scientists. It doesn't happen by accident; first ideas and hypotheses have to be formulated, and then tests emerge. People argued in the past that it was not possible to demonstrate that atoms existed or that neutrinos existed. As the saying goes, "Impossible just means it takes a little longer."

Similarly, string theory can address a number of questions that the Standard Model cannot address, such as the number of families of quarks and leptons, the values of the masses of the quarks and leptons, which phenomena should show the CP violation described in Chapter 8, and whether protons decay. If string theory incorporates quantum theory, gravity, and the Standard Model forces into one description in a consistent way, and if it also explains why there are three families of quarks and leptons and calculates their masses from formulas that have no adjustable parameters, then we will surely have the understanding we hope for. The problem is that the calculations needed to be sure string theory does all of that are very difficult, and they depend on how the small dimensions curl up and on which of many solutions nature actually selected. Thus, even if string theory is indeed right, we won't know it until people are able to do the calculations and confirm that string theory does indeed explain the remaining mysteries. The problem is partly psychological. For example, the development of a quantum theory of electromagnetism was hindered for some time, because a number of calculations of observables gave infinite results when they shouldn't have. Then in 1947 a measurement was reported for one of those observables, and soon thereafter, theorists figured out how to make the calculations finite and meaningful and got the right answer. What changed was that they took it more seriously when there were data, and they had a definite experimental answer to let them know whether the result of their calculation was right. With string theory, the question is whether it will be taken seriously enough so that talented people will focus on the calculations necessary to test the ideas, even when that means investing years in work that might not pan out.

The situation will be similar for the primary theory. If a candidate theory allows us to understand why the rules that describe nature are quantum theory and relativistic invariance, and if it enables us to define space and time in a consistent way, then we will not have much doubt that it is the answer. There may be additional tests there too, though until we get there, we won't know what the tests are. Many of the tests of any idea emerge only after the idea has been formulated and its implications studied (as happened with the tests of the Big Bang, for example, and with the prediction of electromagnetic waves from Maxwell's equations). One can imagine some possible tests of the primary theory. For instance, there is a symmetry that must hold in any relativistically invariant quantum field theory but that need not hold for the primary theory, and one can imagine ways to test whether or not it holds. (Basically, it states that if there is a physical process, then there must be another physical process that gives identical results, obtained by taking the first process and changing all of its particles into their antiparticles, reflecting in a mirror, and running it backward in time.) Once we understand whether and why quantum theory has to be the way nature works, it may turn out that there are necessary modifications to its formulation that appear only at the Planck scale and that can be searched for when we know the associated predictions. Because we can describe some possible tests of the primary theory, it is clear that any arguments that the primary theory is in principle untestable are not valid.

Practical Limits?

Another possible limit to our ability to understand nature may arise because society is unwilling to provide the funding and the commitment to do the basic research needed to test the ideas. As I have argued in this book, particularly in Chapter 5, if there is supersymmetry at the collider scale, then there is a very good chance that the facilities whose upgrading or construction is under way or planned, combined with experiments to study neutrino masses, dark matter, proton decay, and CP violation, and with cosmological data, will provide us with the information needed to formulate and test the primary theory. Even if these facilities are available, however, there are economic risks, because a number of talented and committed people are needed to build and operate these frontier facilities. If the funding comes too slowly, these people will be forced to leave the field, as many had to do when Congress terminated the supercollider project. The people who do this frontier research can be based only at top research universities and a few national laboratories. If those institutions reduce their commitment, there will not be positions for the theorists and experimenters, nor will there be places to train the bright young people who want to learn how and why the universe works or teachers to train and inspire them.

It is interesting to examine the justifications for this funding. In recent years it has been fully understood that a good deal of our economy is based on earlier funding of scientific research. One might think it is the results of research that drive the economy, and they do, but it is not only the results. Fascination with what has been learned from science, and with the way science can lead to understanding our world, initially attracts young people to a career in scientific or engineering research. What they end up working on after they complete their formal education can be very different from what brought them in initially. Many of them see an opportunity to develop products or information technology. The ways that understanding gleaned from science enriches our culture and our view of our relationship to the universe, and the impact of basic research on young people, are probably the two strongest justifications for any society to support basic research generously.

But even apart from these benefits, basic research more than repays society for its cost through the mechanism known as spinoffs. The results of research about Higgs bosons, superpartners, or dark matter are not likely to lead to products that affect the economy. But because these research areas are probing new frontiers, scientists must develop new techniques, and these new techniques invariably lead to new industries. The most recent spectacular example is the World Wide Web, developed at CERN to find new ways to handle data from the LEP collider for international collaborations. As Burton Richter likes to say, if they had named it HEP (for high-energy physics) instead of WWW, or if we had a penny for each use of the web, there would be no funding problem in particle physics.

Another example is accelerators, which were invented to probe more deeply into nuclei and protons and were developed to do further particle physics. They now are used to study materials and matter in many ways, providing knowledge about how to make stronger and safer materials, to determine the structure of viruses, and much, much more. There are thousands of accelerators in use in the world today, and less than a few percent are used for particle physics research. Accelerators have so many medical-related uses that the National Institutes of Health recently increased funding to support not only people who use the accelerators but also the accelerators themselves. The list of spinoffs, and of associated start-up companies that can initially survive because of the guaranteed market provided by particle physics labs and experiments, is very long. It shows convincingly that even apart from the intellectual interest of the research results, funding particle physics and cosmology is an investment that yields large economic returns to society.

ANTHROPIC QUESTIONS AND SUPERSYMMETRY

If the laws of nature were different, could life nevertheless exist? Does the universe have to be a certain way for us to be here? Do the laws of nature imply that a universe (or many) must exist? More precisely, the particles and forces that determine what happens in the universe have various properties and strengths and ranges—if those were different, what would happen? If Planck's constant or the speed of light were different, would the world be different? These questions were first raised in this form in the 1960s and 1970s. It was argued that if any aspect of the laws and constants of nature was much different from what we observe it to be, our universe would be very different and life would not exist. Questions and issues and arguments such as these are called anthropic.

There are several reasons to examine anthropic questions in a book on supersymmetry and its implications. First, we will see that if nature is indeed supersymmetric—or, even more strongly, if nature is described by a supersymmetric string theory—then a number of traditional anthropic arguments are not correct. Second, most or all of the anthropic questions are expected to be addressed by the primary theory. Third, nonscientific and even religious interpretations are increasingly being given to anthropic arguments whose validity is not established. It is worthwhile to question that tendency. To give one example, Vaclav Havel, president of the Czech Republic and a noted writer, has said, "I think the Anthropic Cosmological Principle brings us to an idea perhaps as old as humanity itself: that we are not at all just an accidental anomaly, the microscopic caprice of a tiny particle whirling in the endless depths of the universe. Instead, we are mysteriously connected to the entire universe...."

Let's consider one anthropic argument in some detail. There is an attractive strong force between a neutron and a proton, so they bind into a deuteron, the second nucleus of the periodic table. That happens when a neutron and a proton collide in the sun. Then another proton collides with the deuteron to make a helium–3 nucleus, and some energy is carried off by a photon—the photon is sunlight. Without the deuteron holding the neutron and proton together so that the next collision can occur, it would be very unlikely to produce more sunlight and the helium–3 nucleus that must be there for additional reactions to occur. Thus the attractive force that binds the neutron and proton into the deuteron is essential for the sun to shine. However, from the point of view of the strong force, neutrons and protons essentially behave identically, so there will also be an attractive strong force between two protons. But because the two protons are both electrically charged, they will also feel a repulsive electrical force. In our world that repulsion is sufficient to keep the protons from binding. If the strong force were a little stronger, however, two protons would bind in spite of the repulsive electrical force. Then the reactions that power the sun would proceed at different rates. One can calculate that the sun would burn its fuel far more quickly, and there would not be time for life, dependent on that energy, to evolve on planets. The strong force must be strong enough to bind deuterons, but not strong enough to bind two protons: Its strength must lie in a narrow range.

How should anthropic effects such as this example be interpreted? Some people have chosen to claim that such anthropic effects could not arise from purely natural causes, that it was too improbable that the strong force would have precisely the right strength for life to exist. Such anthropic questions are clearly not ones that can be answered by experiments, but they are nevertheless research questions that can be addressed by the theory. If we had the primary theory, we could work out the answers to such questions. Even assuming only that we will one day have a confirmed string theory, we can address a number of anthropic questions.

For example, to estimate how probable a situation is requires a full understanding of the range of situations that might have occurred. We will see below that in a world described by a supersymmetric unified theory, or a string theory, the probability estimates are far different from what has been claimed.

Anthropic arguments can be split into two kinds. One kind simply takes account of the fact that life exists and concludes that the universe has to be old enough to allow stars and planets to exist and people to evolve. This is not controversial—that people exist is information (data) to include as we try to understand the universe. For example, perhaps lots of universes exist, and only some have the properties that allow people to exist. I will call this kind of anthropic explanation minimal anthropic. I will split anthropic arguments into minimal (as just defined) and nonminimal (all the rest). More subtle distinctions are not needed for our purposes. An example of a nonminimal anthropic argument is the claim that human life would not exist if the strength of the strong force had a value slightly different from its actual value, and that why it has the value it does cannot be explained scientifically, implying it has that value *so that* humans can exist—or even must exist.

One strong reason not to take nonminimal anthropic arguments seriously as implying anything about the world being designed for human life is the past existence of the dinosaurs. The earth was a suitable place for them, and they were a dominant species for about 150 million years (nearly three times longer than mammals, and a hundred times longer than humans). But for a chance asteroid impact 60 million years ago, they would still be the dominant species. Any argument about the meaning of the universe should apply equally to the universe of 100 million years ago and to the universe today. If the universe was planned for humans, somebody got it wrong. Indeed, perhaps one day all humans on earth will be killed by an asteroid impact or by our planet being knocked out of the solar system by the gravitational attraction of a passing planet or star—the probability for such events to happen is not negligible. If that did happen, would it change how the physical universe and its origin should be explained? Nevertheless, let us examine the nonminimal anthropic issues more technically, because they raise interesting issues and suggest questions for the primary theory to answer.

Nearly all scientists expect that scientific explanations will be found for all of the nonminimal anthropic questions. For example, if we did not understand evolution by natural selection, we might think that the ways the human body and mind work are evidence of design or planning—an anthropic explanation. But a century and a half of study has yielded extensive documentation to confirm the evolution of the human eye and brain and body. The scientific, nonanthropic interpretation of the magnitude of the strong force is probably the following one, though before we can be sure, additional calculations will be required.

We have seen that in a supersymmetric Standard Model, and in string theory, the forces of nature are found to be unified. This means that their ratios are fixed by the basic structure of the supersymmetry theory. Then if the strength of the strong force is increased, so must the strength of the electromagnetic force be increased; otherwise, the theory would not be a consistent one. If you increase both forces, deuterons are bound more tightly, but protons also repel more strongly, so the behavior of stars may effectively not change. More important, it is expected that the primary theory is unique: the only possible theory that could describe our world. If that is so, then not only the ratios but also the magnitudes of the forces are fixed, and why they have the values they do is explained scientifically, without reference to whether the universe contains life.

Sometimes proponents of nonminimal anthropic effects try to make the case more compelling by stating it in terms of probabilities. If all the forces (and several other things, such as masses), have some small probability of being in the range needed for human life to occur, people have argued that the way to calculate the probability of all of the effects needed for life occurring simultaneously is to multiply the separate probabilities. Multiplying a number of probabilities produces a *much* smaller probability (for example, one-tenth multiplied six times becomes one-millionth). That is taken to imply that having a universe where life could arise naturally is very, very unlikely.

When that argument was first made three decades ago, it was worth examining. But it would be correct to multiply the probabilities only if the separate ingredients were independent. For example, if one force strength being a certain size implied that another was some definite size, the second should not be included in a probability calculation. Now we understand that in a world described by supersymmetry, the forces are unified, so if one force is in the right range, the others must also be. The probabilities are not independent and should not be multiplied. If the world is described by a string theory, not only the ratios of the forces are fixed but also their magnitudes.

Another nonminimal anthropic effect is that neutrons must be a little heavier than protons if the synthesis of heavier elements is to proceed in a way consistent with having life exist. That translates into requiring up quarks to be lighter than down quarks. Those quark masses are calculable in string theory, though they have not yet been calculated. If the world is described by a unique string theory, this outcome for the quark masses will not be independent of the others. Thus even if someone wants to claim that some aspects of the universe that we do not yet understand could be nonminimal anthropic ones, their probabilities should not be multiplied if they are correlated in the theory. Hence the combined probability would not be nearly so small as is usually claimed—in fact, it is not small at all.

Until we have the primary theory and understand how to calculate its implications, we cannot settle all anthropic questions, so they are valid issues to study. In spite of this, and in spite of the obvious interest of anthropic issues, most physicists do not study them, because they expect there will finally be few if any anthropic effects that will not be better understood by normal physics. There are, however, two very weak senses in which many more physicists expect anthropic effects to occur. One is simply the observation that our universe must have properties consistent with the emergence of human life; that is, it must be minimal anthropic. In other words, this line of reasoning simply takes the data into account and reads into them no implications for how we give meaning to life and the universe. For example, life will obviously not evolve until stars and planets have formed. In addition, heavy elements are essential for life, and we know that the heavy elements are made in exploding stars (supernovas). Therefore, the universe has to be old enough for the first generation of stars to form, grow old, and die, and for a second generation of stars to form with planets, before life can evolve.

The second sense in which we can anticipate anthropic effects is more challenging. Although we do not yet understand how our universe actually originated, attractive ideas have emerged that provide mechanisms for universes to arise from nothing at all. These ideas suggest that universes first occur as tiny, Planck scale entities and then inflate in a tiny fraction of a second to sizes large enough to see (were someone there to watch). Their total energy, including gravitational attraction, is zero, but a large part is in potential energy, and that is released in the form of particles—an event we call the Big Bang. These ideas imply that new universes are being created by such processes randomly and all the time. It may be that the laws of nature are different in different universes. Most people working on such questions think that there will be only one possible primary theory and that it will govern all the universes, so the laws themselves will have to be the same. But perhaps the constants that enter into the laws—such as force strengths, Planck's constant h, the speed of light c, the gravitational force strength G, and the cosmological constant (it is really the ratios of such constants that should be considered, but we will not concern ourselves with that)—can be different in different universes. Then life may emerge in some universes but not in others. If there are very many universes, then some, perhaps many, will have the right conditions for life to arise.

THE END OF SCIENCE?

Suppose that eventually we do figure out what the primary theory is, and suppose we are able to test it so well that most of us are convinced that we indeed understand (scientifically, without supernatural influences) how our universe works, why it is that way, and why some universes have properties consistent with the emergence of life. What does that imply? Although humans began to wonder about the universe perhaps 40,000 years ago, and although they started a more systematic quest to understand it some 2,600 years ago in Greece, until about 1600—a mere 400 years ago—there was only a little descriptive progress. No aspect of how nature actually works was understood. Then modern science dawned, and today we have started research on the primary theory itself. Before the 1970s, we did not know how the world worked; now, to a large extent, we do. Perhaps in a few decades or less, we will reach the end of this quest. Our universe has properties consistent with having humans in it, though it is indifferent to whether they are there and to whether they understand it. For me, and I hope for many people, achieving an understanding of why there is a universe and why it is the way it is will be a source of immense pride and dignity and meaning in the face of that indifference.

If we achieve that understanding, it will also be a source of great sadness for some people, because a long journey that added meaning to the lives of many will have ended. The sadness is for the generations yet to come, who will not be able to share the excitement of the search for understanding of basic aspects of the natural world. Many people, many of them scientists, have said that we will not reach that end—that there will always be new questions and that each discovery will lead to additional questions. But why should that be so? There is no known reason why the quest should not end. Sometimes the analogy with exploration of the surface of the planet earth is helpful: For many centuries there was more to explore, and then one day the mapping was essentially complete. Of course, there have already been a few occasions when someone said there was nothing new to learn, but if you actually examine the concerns of leading scientists since the 1860s, you find that active scientists always knew this claim was wrong. However, the situation truly is different today. We have extended our investigations back to the beginning of the universe, out to the edge of the universe, and down to the fundamental constituents of matter. This doesn't guarantee that we will achieve total understanding, but it does demonstrate that the historical analogies need not be relevant.

Yes, it is possible that the scientific pursuit of more fundamental effective theories could end, not because we *couldn't* get all the way to the primary theory, but because we *did*. That wouldn't mean that science itself had ended. Although we know the equations and the law(s), we would still be far from knowing all the solutions. All areas of science except particle physics and cosmology are open-ended. And if we indeed discover the primary theory, it will change how we view life and meaning itself.

The Standard Model Higgs Mechanism

This appendix explains in more detail how the Standard Model Higgs mechanism works. It contains an equation and a slightly technical argument, but anyone willing to take one step at a time can follow it—no knowledge of mathematics is needed. This account will give the reader a better sense of why the Higgs physics of the Standard Model seems ad hoc and odd, though it works well technically. The next appendix shows how these results can be derived in a supersymmetric world so that they seem natural and inevitable.

The universe is full of fields. Every mass sets up a gravitational field, and every electric charge an electric field. All the particles are quanta of fields. All of these fields carry energy and add an energy density throughout space. For all of them, the state of lowest energy density is the state where the fields are zero. This is important because any system will end up in its lowest energy state, and the universe is no different, so apart from the fields arising from the particles, the lowest energy state of the universe is basically empty. Actually, this statement is not quite correct because of the uncertainty principle, and possibly because of cosmological energy densities, but we can ignore those issues for this discussion. We speak of the lowest energy state of the universe as the vacuum, the normal background state of the universe.

In the Standard Model we add a Higgs field in an *ad hoc* way, without worrying about its source (in the sense that massive particles are a source of a gravitational field, electrically charged particles are sources for electromagnetic fields, and so on). The Higgs field (let's call it *h*) increases the energy density of the universe, adding an amount of energy we can call *E*. The relation between the Higgs field, *h*, and the amount of energy it adds, *E*, is assumed to be given by the equation

$$E = M^2h^2 + Ah^4$$

where *A* is a positive but unknown constant (all we will need to know about *A* is that it is positive). The units of the quantities in this equation won't affect our analysis, so we'll ignore them. Usually M^2 would be interpreted in this equation as the square of the mass of the quanta of the Higgs field. If *A* and M^2 were known, we could use this equation to calculate the amount of energy density, *E*, added for a given size of *h*. Even though we do not know M^2 and *A* in the Standard Model, we can learn several important results simply by looking at how this relation behaves.

Let's sketch a graph of how *E* behaves as *h* increases from zero to larger values. For *h* = 0, both terms on the right are zero, so *E* is zero. This fixes the point at *E* = *h* = 0. Next let *h* take on a small value; for example, think of *h* = 0.1. So long as M^2 and *A* are positive (whatever their actual values), *E* is positive, and as *h* increases, *E* gets larger, so it looks like the sketch in Figure A.1.

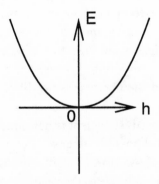

FIGURE A.1.

The coefficient that we are calling M^2 was given that name instead of some name such as "B" because in the theory, that coefficient usually would be interpreted as the square of the mass M of the quanta of the Higgs fields, the Higgs bosons. With this interpretation, E should indeed be zero for $h = 0$ and should get larger as h does. However, we take a dramatic step for the Higgs field and assume instead that M^2 is a *negative* quantity. At first this may not seem to make sense for the square of a mass, but let's examine the consequences for the way E behaves, which turn out to be remarkable. For $h = 0$, E is again zero. The first term is always negative because M^2 is negative, and the second term is always positive. Now let h have a small value such as 0.1. If A and M^2 are not too different in size, the first term is larger because $h^2 = 0.01$ but $h^4 = 0.0001$ (if A happens to be very large, pick a smaller h), so the value of E always starts out negative as h increases from zero. As h gets larger, eventually E becomes positive because the second term dominates, so the graph of E looks like Figure A.2. We don't know exactly how to draw it—for example, how deep it goes, or just where it becomes positive—but the basic shape has to be like what is shown, and that is all we need to know.

If you think of this figure as a kind of energy landscape of the universe, much like a valley with a ball bouncing around, the ball and the universe will both eventually settle at the bottom, in the state of lowest energy. We see in this case that the state of lowest energy does not occur for a zero Higgs field, but rather for a value of h different from zero (the precise value does not matter here). The universe is filled with a nonzero Higgs field in its vacuum state!

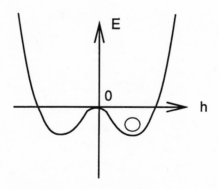

FIGURE A.2.

Just as most particles typically interact with one another, the Standard Model particles can interact with the Higgs field. Only the photon, neutrinos, and the gluons do not directly interact with it. Interacting with the Higgs field that fills the vacuum is a little like trying to walk through water instead of air—you feel heavier. That is, you behave as though you have mass, which for particles is the same as having mass. The interaction with the Higgs field has allowed the particles to have mass in the theory in a consistent way. All of the quarks and the charged leptons and W and Z get mass from the interaction with the Higgs field.

The Higgs physics of the Standard Model consists of three parts. First, one has to postulate the existence of the Higgs field. Second, one has to assume the above relation between the energy density due to the Higgs field and the size of the field itself, with M^2 a negative quantity. Note that M^2 negative is what triggers the Higgs mechanism; that will be the key clue we look for in the next appendix when we look at how the Higgs mechanism can be derived in a supersymmetry theory. Third, as discussed in Chapter 7, there will be quanta of the Higgs field—Higgs bosons—and they will be observable, which is the crucial test of the whole idea. The way supersymmetry improves on the Standard Model Higgs physics is explained in Appendix B.

The Supersymmetry Explanation of the Higgs Mechanism

In Appendix A we saw that the essential feature needed for the ground state of the universe to be filled with a Higgs field was a rather unnatural condition: A quantity called M^2 that would usually be positive had to be forced to be negative. For the Standard Model, that condition had to be imposed by hand, rather than emerging from the theory as one would prefer.

The key to understanding the way the negative M^2 does emerge in the supersymmetry theory, rather than having to be imposed, is to realize that the supersymmetry theory is naturally formulated near the Planck scale, where the forces unify and the theory is simpler, as we have seen in this book. At that scale we do not yet know all of the masses and other relevant quantities. But M^2 and the other quantities have normal positive values—the precise values don't matter. M^2 is not negative and need not have any special value. To understand the Higgs physics, it is essential to think in terms of the effective theory approach of Chapter 3.

We need to know the value of M^2 (and of the other quantities too) at the scale where the Standard Model is the effective theory if we are to understand the world at that scale. Because the supersymmetry theory is a quantum field theory, we can use it to calculate the values of interaction strengths and masses at one scale if we know them at another. Thus we can calculate how the value of M^2 changes as it is viewed at increasingly larger scales, from the Planck scale to the Standard Model scale. This property of quantum field theories—that we can calculate how quantities such as M^2 change at different scales—is not obvious. Four decades elapsed after the discovery of quantum theory before this property was recognized, and longer before it was fully understood. It is the same procedure that allows us to calculate how the strengths of the forces change at different scales and approach the same strength near the Planck scale, as described in Chapter 4, so we have good reason to take the method very seriously. The actual calculation is rather technical, but the resulting behavior is illustrated in Figure B.1, which shows squares of masses on the vertical scale and energy horizontally, so the Planck scale corresponds to a large energy compared to the Standard Model scale. We could show other masses too, but for simplicity we look only at the one we need, M^2.

We see that as we go from the Planck scale to the scale of the Standard Model, M^2 starts out positive and decreases, becoming negative as we approach the Standard Model scale. If the value at the Planck scale were a little different, it would become negative anyway, just at a somewhat different place. Until M^2 becomes negative, the W bosons, Z bosons, quarks, and leptons are massless, because the minimum energy of the universe occurs with no Higgs field present to interact with, as in Figure A.1. Thus this approach

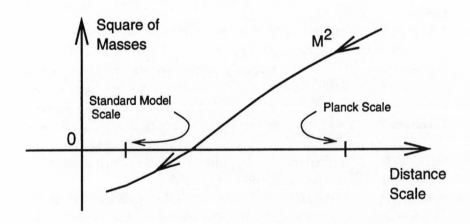

FIGURE B.1.

implies that they were massless during the Big Bang (which took place at tiny distances or large energies on the order of the Planck scale), and shortly after, until the universe had cooled to energies typical of the Standard Model and collider energies (a thousandth of a second or so). Then the universe was in a lower-energy state, as in Figure A.2, with a nonzero Higgs field present that particles interacted with to become massive.

There are two major tests of whether this is indeed the correct understanding of the Higgs physics. One is already successful, which gives us a lot of confidence in this derivation. To explain the Higgs physics, the value of M^2 must become negative at a distance scale smaller than the Standard Model scale itself, so the particles have masses at the Standard Model scale. The equation that determines the scale at which M^2 becomes negative (where the curve in Figure B.1 crosses the axis) contains one quantity that is not fixed by the supersymmetry theory—namely, the strength of the interaction of the top quark with the Higgs field. In order for M^2 to become negative in a relevant way, the top quark interaction strength must be about the same size as the strength of the interaction of the W and Z bosons with the Higgs field. That may sound reasonable, but all of the other quarks and leptons have much smaller interactions with the Higgs field, so it was a surprising requirement when it was first noticed (in 1982). At that time the top quark had not yet been observed, so this was a prediction, and it was unexpected. For example, many experiments were constructed to search for the top quark under conditions in which it could not be found if this prediction were true. In the 1990s, as the top quark was finally observed, first indirectly and then directly at Fermilab, the prediction was indeed seen to be true, a result that greatly increased our confidence that supersymmetry does indeed explain the Higgs physics.

The second test also involves a prediction, and an extremely important one. The Higgs physics relates M^2 and other quantities we do not know—quantities characteristic of the basic theory at or near the Planck scale—to quantities we have already measured, such as the masses of the Z and W bosons. The quantities we do not know include the masses of the superpartners. Thus we can write an equation with the mass of (say) the Z boson on one side and masses of superpartners on the other side. For such an equation to be meaningful, the terms on one side should not be much different in size from the terms on the other. For example, we wouldn't trust equations such as $a = b - c$ if a were about 2 but b and c were about 100,000, because it is unlikely that we could calculate or somehow know b and c to such an accuracy. Therefore, the superpartner masses cannot be very much larger than the Z boson mass if this whole approach is valid. This is the only place where we can use the theory to relate the unknown superpartner masses to known masses, so on the one hand, it is a major test of the correctness of the supersymmetry explanation of the Higgs physics, and on the other, it is the most significant reason why we expect the masses of the superpartners to have values that allow them to be produced at Fermilab or even LEP. This connection also suggests that if the superpartner masses are much larger than the Z boson mass, then the apparent success of the supersymmetry theory in explaining the origin of the Higgs physics of the Standard Model could be an accident.

Charginos and Neutralinos

The superpartners that weigh the least are the ones most likely to be produced first, because it takes less energy to produce them (or, for a given amount of energy, more of them can be produced, and therefore they can be more easily detected). We don't know for sure which are the lightest ones, but most approaches suggest they might be the photino, Wino, Zino, and higgsinos. Perhaps these superpartners will be found soon, and there will be exciting reports about their discovery.

There is a subtle and unfamiliar effect, however, that implies that the superpartners you read about will indeed be these but that they will be called charginos and neutralinos instead of photinos, etc. The effect is not unique to supersymmetry but is, rather, a general property of many physical systems. Readers who want to relate what is reported to the particles in this book should be aware of this complication.

Ideally we would characterize all of our particles by giving a list of their properties, such as mass, electric charge, weak charge, and spin. Nature sometimes allows two or more different particles to share some of the properties in this list. However, when two or more particles share some properties, and the other properties are not conserved ones, the particles involved can "mix" with one another. In particular, particles with the same electric charge and spin can do so, because the interaction with the Higgs field (which determines how particles get mass) does not preserve the weak charge. The particles that are produced in experiments are always those with definite mass, even if they are mixtures of particles with different weak charges.

For example, the superpartners of the W boson and the electrically charged Higgs boson are affected this way. These two superpartners can mix, so the states of definite mass that are produced are not Winos or charged higgsinos, but mixtures, which we call *charginos*. The distinction between charginos and Winos is of technical interest to physicists trying to interpret data about supersymmetry, but it is of little interest to anyone else.

Another set of particles that mix is the photino, the Zino, and the two electrically neutral higgsinos. We call the combinations of these particles that have definite mass and thus can be produced in experiments *neutralinos*. Just as for charginos, the distinctions between the states of definite mass and those of definite weak charge are only of technical interest, with one exception. The lightest of the neutralinos is likely to be the lightest superpartner, so it is a very good candidate for the cold dark matter of the universe, as discussed in Chapter 6. The mixing phenomenon implies that the lightest neutralino is likely to be a combination of higgsinos, photino, and Zino, so its properties will be harder to establish than if it were purely one or another type. The mixing not only adds complexity to the terminology; it also complicates the effort to learn the implications of superpartner data for the dark matter.

Extra Dimensions—
Large Extra Dimensions?

When we achieve a true understanding of nature, even familiar ideas such as space and time will have to be derived, as will the number of dimensions in which we live. Serious scientific thinking about why we seem to live in three space dimensions arose surprisingly long ago. The first person to publish on the question may have been Paul Ehrenfest. He pointed out in 1917 that the equations describing the motion of planets around the sun, and the similar equations for electrons bound to nuclei in atoms, only had stable solutions in three space dimensions. This seems to have been the first anthropic argument: Atoms and the solar system can retain their structure long enough for us to be here only if we live in three space dimensions. People's awareness of the question of how many dimensions we lived in was probably stimulated by Einstein's special relativity, which related space and time (1905), and even more by Einstein's general relativity (1915), which used mathematics valid in a general number of dimensions.

In 1919 Theodor Kaluza, and independently, a few years later, Oskar Klein, pointed out an astonishing result. Consider a world with four space dimensions, one time dimension (5D), and only one force, gravity. This is a sensible world to examine because gravity is a universal force and thus affects everything. Write the equations that would form Einstein's general relativity theory in that world. Next assume that one of the dimensions is much smaller than the others, so we are not aware of it. Just as every point on the floor of a room has a third vertical dimension, so every point in the 4D world could have an associated fifth dimension, but one so small that we can't tell we are moving in it. Then look at the equations that define the theory. Kaluza and Klein showed that the equations split up into two sets. One was identical to Einstein's equations for gravity in the 4D world. The other was identical to Maxwell's equations for the electromagnetic force! The component of the gravitational field in the direction of the tiny dimension obeyed the same equation as the electromagnetic field and could be identified with it. A quantity that behaved like electric charge could be defined. Perhaps electromagnetism could be understood if we really lived in a 5D world that included one tiny dimension. The gravitational force in the tiny direction was our electromagnetism.

At that time, our understanding of the particles and forces was not adequate to allow further progress with these ideas, and no major advances occurred involving extra dimensions for over half a century, but the Kaluza-Klein ideas were so remarkable that most theorists interested in such questions learned about them. In the 1980s string theories became exciting approaches, and they required for consistency that the particles be strings that vibrated in nine space dimensions and one time dimension. That turned out to be just right for the forces too. A world with only a gravitational force in 10D, when looked at as a world with three large space dimensions and six tiny ones, behaved as one with forces that can be interpreted at large distances as the 4D gravitational force, plus electromagnetic and weak and strong forces—no others! In a sense, the extra dimensions are manifest as forces when they become tiny. How all of this works is still very much RIP.

How do we see the effects of the extra dimensions if we cannot probe them directly because they are too small? There are actually a number of ways in which they may indirectly affect our world and thus enable us to deduce their presence. We just saw that perhaps they can explain why particles in our world interact the way they do and why the electromagnetic and weak and strong forces exist and have the properties they do. The particles themselves are strings in 10D in this view, and different particles correspond to different behavior of the strings, so we can relate the extra dimensions to different amounts of electric charge and other charges. Cosmology—which consists of the equations that describe the behavior of the universe—is different for a 4D world or a 4D part of a higher-dimensional world; for example, the way forces vary with distance depends on the number of dimensions. Possibly the mechanism that keeps some dimensions small also affects the relationship between bosons and fermions and thus is related to how the supersymmetry is hidden in our 4D world. The number of compact dimensions and the way they are curled up is expected to be related to the number of families. Thus there are many pieces of experimental information that can be related to the number of dimensions, once there is a theoretical framework to allow us to study these questions.

The reader may have seen somewhat confusing statements about the number of extra dimensions—a total sometimes of ten and sometimes of eleven. That is because it was learned in 1995 that until then the role of one dimension that is required by the theory for consistency had not been understood, because the string theory equations were formulated and studied approximately. The role of the missed dimension was different from that of the other dimensions. That is, strings still vibrate in nine space dimensions, but in the complete theory, strings themselves have a kind of thickness in another dimension. These kinds of developments and modifications are what one expects in any new field—again, it is all very much RIP.

The connection of the extra dimensions with supersymmetry is not yet well understood. String theories must be supersymmetric and have extra dimensions to be consistent, but the implications of this for how supersymmetry is hidden are not yet known. Perhaps most important, predictions for the masses and interactions of the superpartners also reflect the presence of the extra dimensions and how they are curled up. Thus experimental data on the superpartners may play a crucial role in our coming to understand the extra dimensions.

Large Extra Dimensions

In the 1990s theorists began to wonder what arguments, theoretical or experimental, could tell us how large the extra dimensions might be. Until then everyone assumed they were all about the size of the Planck scale, and that is still the most likely answer, but as always, it is important to examine the evidence. People had thought that a number of the successes of the theory would no longer hold if extra dimensions were much larger than the Planck scale, but under certain conditions that turned out not to be the case. There are many alternatives to study—for example, perhaps some forces (such as gravity) can act in larger extra dimensions, but not the other forces, or perhaps all the forces can. There are many interesting consequences that allow experimental study—for example, the gravitational force has never been studied in experiments at distances smaller than about a millimeter, and perhaps it behaves differently in that domain than at larger distances (this is precisely the kind of situation discussed in Chapter 1 for extending theories into new domains). Perhaps in collisions of particles gravitons can be emitted into the extra dimensions and not appear in detectors, leaving a new kind of experimental signature. Perhaps the world is 4D down to 10^{-31} meter, then 5D or 6D down to 10^{-33} meter, and 10D at the Planck scale. All of these extremely interesting alternatives are currently being studied—RIP.

Even if no experimental trace of extra dimensions much larger than the Planck scale is found, thinking about large extra dimensions will have been very beneficial. Before this work, most theorists who think about these issues had become convinced that extra dimensions existed, but they were often thought of more in a mathematical than in a physical way. Now that the reality of extra dimensions is being examined in more tangible ways, a number of people are gaining intuition and experience that will be valuable in ongoing efforts to improve our understanding of the theory and how it connects to the real world.

If extra dimensions that are extremely large compared to the Planck scale (say, a billion or more times larger than the Planck scale) turn out to exist, then some features discussed in this book would need to be reexamined. Much would not change—supersymmetry would still be a basic feature of the underlying string theory—but the implications could change because the extra dimensions would affect our world.

Some Recommended Reading

2000

The past two decades have seen the appearance of a number of very good books and articles about science for general readers, covering both what has been understood and RIP. I list here a few that may be of particular interest to readers of this book.

On the main subject, supersymmetry and its implications, there is little yet available that is worthwhile and nontechnical. Some articles may be helpful: "Supergravity and the Unification of the Laws of Physics," Daniel Z. Freedman and Peter van Nieuwenhuizen, *Scientific American*, February 1978; "Is Nature Supersymmetric?" Howard E. Haber and Gordon L. Kane, *Scientific American*, June 1986; and "Desperately Seeking SUSY," Hans Christian Von Baeyer, *The Sciences*, September/October 1998. Considerable discussion of the early history can be found in the Yuri Golfand Memorial Volume *Many Faces of the Superworld*, ed. M. Shifman, Singapore: World Scientific, 1999.

Readers who would like to learn more about the Standard Model of particle physics can turn to my earlier nontechnical book *The Particle Garden*, Reading, MA: Addison-Wesley (Helix Books), 1995. Some people with scientific training may want a somewhat more technical treatment of the Standard Model, which they can obtain from *Modern Elementary Particle Physics*, Gordon L. Kane, Reading, MA: Addison-Wesley (updated edition 1993), a book at senior undergraduate level. Two nice nontechnical studies of the history of the Standard Model, with considerable information about the experimental foundations, are *The Second Creation*, Robert Crease and Charles Mann, Rutgers University Press, 1996; and *The Rise of the Standard*

Model, ed. Lillian Hoddeson, Laurie Brown, Michael Riordan, and Max Dresden, Cambridge, England: Cambridge University Press, 1997.

There are surprisingly few nontechnical places to turn simply to learn about quantum theory. One satisfactory treatment appears in the relevant chapters of *Physics for Poets*, 4th ed., Robert March, New York: McGraw-Hill, 1996. A book that focuses on the interpretation of quantum theory (although the technical formulation of quantum theory is not in doubt, how we interpret it is still an active area of study) is *Where Does the Weirdness Go?*, David Lindley, New York: Basic Books, 1996; and a nice treatment of the interpretation of quantum theory is given in the relevant chapters of *The Quark and the Jaguar*, Murray Gell-Mann, New York: Freeman, 1994.

Several recent books focus on cosmology, including its connections to particle physics: *A Brief History of Time*, Stephen Hawking, updated 10th anniversary edition, New York: Bantam Books, 1998; *Before the Beginning*, Martin J. Rees, Helix Books, 1997; *The Inflationary Universe*, Alan H. Guth, Reading, MA: Addison-Wesley (Helix Books), 1997; and *The Whole Shebang*, Timothy Ferris, Simon & Schuster, 1997. *Blind Watchers of the Sky*, Rocky Kolb, Reading, MA: Addison-Wesley (Helix Books), 1996, describes the history of cosmology, and *The Five Ages of the Universe*, Fred Adams and Greg Laughlin, New York: The Free Press, 1999, describes the future of the universe.

Some books and articles about string theory and extra dimensions are "The Theory Formerly Known as String Theory," Michael Duff, *Scientific American*, February 1998; *Beyond Einstein*, Michio Kaku and Jennifer Thompson, New York: Oxford University Press, 1995; *Hyperspace*, Michio Kaku, New York: Oxford University Press, 1994; and *The Elegant Universe*, Brian Greene, New York: Norton, 1999.

Two books that speculate in a general way about learning the primary theory are *Dreams of a Final Theory*, Steven Weinberg, New York: Pantheon, 1992; and *Theories of Everything*, John D. Barrow, Oxford, England: Clarendon Press, 1991.

Glossary

SYMBOLS

e; $\tilde{e}$	electron; selectron
μ; $\tilde{\mu}$,	muon; smuon
τ; $\tilde{\tau}$	tau; stau
ν; $\tilde{\nu}$	neutrino; sneutrino (There are separate sneutrinos for electron, muon, and tau, which are sometimes written with subscripts.)
u, d, c, s, t, b	quarks (up, down, charm, strange, top, bottom)
$\tilde{u}$, $\tilde{d}$, $\tilde{c}$, $\tilde{s}$, $\tilde{t}$, $\tilde{b}$	squarks [up squark, down squark, charm squark, strange squark, top squark (sometimes "stop"), bottom squark ("sbottom")]
W; $\tilde{W}$	W boson; Wino
Z; $\tilde{Z}$	Z boson; Zino
γ; $\tilde{\gamma}$	photon; photino
g; $\tilde{g}$	gluon; gluino
G	Newton's constant, which determines the strength of the gravitational force
h	Planck's constant, which determines the size of quanta of energy and of quantum theory effects in general
c	Einstein's constant, representing the speed of light in vacuum (or the charm quark, depending on the context)
Powers of 10	10^{-9} = one billionth, 10^{-6} = one millionth, 10^{-2} = 1/100 (one hundredth), 10^{0} = 1, 10^{1} = 10, 10^{2} = 100, 10^{3} = 1000, 10^{6} = one million, 10^{9} = one billion

ACRONYMS AND ABBREVIATIONS

Fermilab   Fermi National Accelerator Laboratory, Chicago, IL
SLAC   Stanford Linear Accelerator Center, Palo Alto, CA
CERN   Centre Européan Recherche Nucléaire, Geneva, Switzerland
LEP   Large Electron–Positron Collider, at CERN
SLC   Linear electron–positron collider, at SLAC
LHC   Planned large proton–proton collider at CERN, intended to
           take data in 2005
NLC   The next linear collider. The possibility of proposing such a
           collider is under study by several countries and laboratories.
           There are discussions about making it an international facility;
           it would be an electron–positron collider with energy four or
           more times that of the SLAC SLC.
FMC   First muon collider
CDF   Detector at Fermilab
D0   Detector at Fermilab
RIP   Research in progress
LSP   Lightest superpartner

TERMS

Accelerator

Accelerators are machines that use electric fields to accelerate electrically charged particles (electrons, protons, and their antiparticles) to higher energies. If accelerators are linear, they need to be very long to achieve the desired energies, so in some, magnets are used to bend the particles around and back to the starting point, giving them a little extra energy each time around.

Antiparticle

Every particle has an associated antiparticle, another particle with the same mass but all charges opposite. If a particle has no charges, such as the photon, it is its own antiparticle. Often the antiparticle is denoted by writing a bar over the particle's name; hence, for example, the electron antiparticle (also called the positron) is denoted $\bar{e}$.

Atom

An atom has a nucleus surrounded by electrons bound together by the electromagnetic force. Ninety-two different atoms occur naturally, making ninety-two different chemical elements, with nuclei having one to ninety-two protons. The atoms are electrically neutral. The diameter of an atom is about 10,000 times larger than the diameter of its nucleus.

b-factory

A b-factory is a facility designed to produce and detect large numbers of b-quarks, at least 100 million a year. Planned b-factories are electron–positron colliders, but a proton collider could also be used if an appropriate detector could be made. The main goal of b-factories is to study CP violation.

Baryon

A baryon is a composite particle made of three quarks, any three of the six. Protons and neutrons are baryons.

Baryon Asymmetry

See *Matter Asymmetry*.

Beams

One way to learn more about particles is to cause them to collide with one another and see what happens. Beams of electrons and protons can be made by knocking apart hydrogen atoms and applying electric fields. Positrons and antiprotons don't exist naturally, because they annihilate as soon as they encounter an electron or proton; they can be made by hitting a target with energetic protons or electrons and then collected by placing magnets after the target, arranged so as to bend each kind of particle in a different path. Then bunches of them are accelerated to higher energies. When a particle hits a target, every kind of particle is made with a certain probability, so other beams of particles can also be made (neutrons, muons, kaons, neutrinos, etc.) by the judicious arrangement of magnets and material.

Big Bang

Several strong kinds of evidence imply that our universe began as a tiny, dense gas of particles that has been expanding since. That is, our universe began in a "hot Big Bang."

Black Hole

Black holes form whenever sufficient matter is packed into such a small space that the resulting gravitational force at the surface is strong enough to prevent anything, including light, from escaping. They can be microscopic in size or formed from billions of stars. They can be detected from their indirect effects on nearby matter. The theoretical study of their properties can greatly clarify our understanding of basic questions.

Boson

Bosons are any particles that carry an integer unit of spin $(0, 1, \ldots)$. They have different properties from particles with half a unit of spin (fermions). In particle physics, *boson* has a more specific meaning: Bosons (photons, gluons, and W and Z bosons) are particles that are the quanta of the electromagnetic, strong, and weak fields. They transmit the effects of the forces between quarks, leptons, and themselves. Higgs bosons are quanta of a Higgs field; as this book is being written, the evidence for the existence of Higgs bosons is still indirect.

Charge—Electric, color, weak

Each particle can carry several kinds of charges that determine how it interacts with others. Electric charge is familiar to us in everyday life. Particles can have positive or negative electric charge, or none. Color charge and weak charge are not familiar because their effects can be felt only at distances smaller than the size of a nucleus. Color charge and strong charge are the same thing. A particle cannot have random amounts of charge; only certain discrete amounts are allowed. The extent to which a particle feels each force is proportional to its associated charge. Quarks and gluons carry color charge; quarks, leptons, and W and Z bosons carry weak charge.

Chemical Elements

Ninety-two different stable or long-lived nuclei can be formed from neutrons and protons bound together. Each forms atoms by binding as many electrons to the nucleus as it has protons (so the nucleus is electrically neutral), giving ninety-two different atoms. These atoms are the smallest recognizable units of the ninety-two chemical elements.

Cold Dark Matter

Particle physics theories that extend the Standard Model often predict the existence of new, stable particles that were present in the early universe and survive today, making up a large fraction of the matter of the universe. These particles interact weakly and they are usually massive, so they move slowly—they are cold. An example of such a particle is the lightest supersymmetric partner. Astronomers have evidence from the motion of galaxies, and from the large-scale structure of the universe, that cold dark matter exists.

Collider

A collider is made by accelerating beams of particles and causing them to collide. The energy of the colliding beams can provide much more energy (that can be used to make new particles) than if the beams hit stationary targets. Two challenges confront colliders: getting to larger energies and getting to higher intensities.

Color Charge

See *Charge*. Color charge and strong charge are the same thing.

Color Field

Any particle carrying color charge (or strong charge) has an associated color field (or strong field) around it. Any other particle carrying color charge feels that field and interacts with the first particle.

Color Force

The force between two particles carrying color charge. The color force (or strong force) binds quarks into protons and neutrons. The residual color force outside protons and neutrons is the nuclear force that binds protons and neutrons into nuclei. The color force is mediated by the exchange of gluons.

Composite

Any object made of other objects is composite, as are atoms, nuclei, and protons. If quarks and leptons had followed the historical trend that matter at each level turned out to be composites of smaller constituents, experiments should already have shown evidence of their compositeness.

This, combined with theoretical arguments, strongly suggests that quarks and leptons may be the ultimate constituents of matter—the indivisible "atoms" of the Greeks.

Constituents

Any objects that are bound together to make larger objects. For example, atoms are constituents of molecules, nuclei are constituents of atoms, and so on. See also *Composite.*

Cosmic Rays

Protons and some nuclei that are ejected from stars, especially supernova explosions, move throughout all space. These "cosmic rays" impinge on the earth from all directions. They normally collide with nuclei of atoms in the atmosphere, producing more "secondary" particles, mainly electrons, muons, pions, etc. A number of cosmic ray particles go through each of us every second, and they can interact in detectors and mimic signals of previously undetected particles, so experimental equipment must shield against them or be able to recognize them so that they can be discounted as signals of new physics.

Cosmology

Cosmology is the study of the universe as a whole, its properties, and its origin.

Cosmological Constant

The name given to a term that may be found to occur in the equations that describe the universe. If it is not zero, this implies there is a force that is slowly increasing the expansion rate of the universe. There are two puzzles about the cosmological constant: The observed value seems to be far smaller than estimates would imply, and recent data suggest that it is not exactly zero, so an explanation is needed for why it has a particular nonzero value.

CP Violation

Interactions of quarks, leptons, and bosons are normally invariant under a symmetry operation called CP, the combined operations of "parity" and "charge conjugation." A small violation of this invariance is observed, which may have important implications and be an important clue to deeper understanding of how nature works.

Dark Matter

Particle physics theories that extend the Standard Theory predict several forms of matter that may exist in large quantities throughout the universe and make up most of the matter of the universe. Some move slowly and are called cold dark matter; others move rapidly and are called hot dark matter. Study of the motions of galaxies and of the formation of clusters of galaxies suggests that such dark matter exists, as do theoretical cosmological arguments based on other data. See also *Cold Dark Matter* and *Hot Dark Matter*.

Decay

The quarks and leptons and bosons that are the particles of the Standard Theory have interactions that allow them to make transitions into one another. Whenever one of them can make transitions into lighter ones, that transition will occur with a certain probability, and we say the heavier one is unstable and has decayed into the lighter ones. Decays are really transitions—the final particles were not contained in the initial particle. The initial particle disappears, and the final ones are created. In the Standard Model the up quark, the electron, and the neutrinos do not decay; the other fermions and the W and Z do decay. All of the superpartners except the lightest superpartner are expected to decay.

Detector

The properties of particles and their interactions are studied by observing their interactions and decays. These observations are done with detectors, which can be thought of as cameras that record information in several ways, not only on film. In order to obtain new results at the forefront of today's inquiry into the nature of matter, detectors have to be very large, and they often require the development of better technologies. Every particle physics experiment has one or more detectors.

Deuteron

A deuteron is the second heaviest nucleus (after the lightest, hydrogen, which has a single proton). It is composed of a neutron and a proton bound together by the nuclear force. The deuterium atom has a single electron bound to a deuteron (one electron because there is one proton).

Dirac Equation

The Dirac equation incorporates the requirements of both quantum theory and special relativity in its description of the behavior of fermions. It requires that fermions have the property called spin, and it predicts the existence of antiparticles. It was written by Paul Dirac in 1928.

Effective Theory

Each part of the physical world can be described by a sub-theory that applies over a certain distance scale or energy scale. Such sub-theories are called effective theories. Explanations in a given effective theory can ignore much of the rest of the world, which has effects on the part of interest through a few inputs or parameters. Every part of our description of the physical world is an effective theory, except the ultimate theory that I call the primary theory. See Chapter 3.

Electric Charge

See *Charge.*

Electron

A fundamental particle. The electron has one unit of negative electric charge and half a unit of spin. It is a fermion.

Electron Collider

Short for electron–positron collider. One important way to study particle interactions and search for new particles is to accelerate an electron and a positron to high energies and then collide them, using a detector to study what emerges. The energy to which they are accelerated is chosen to fit the question of interest. For example, to study CP violation in b-quark decays, the energy is chosen to maximize the production of b's in an appropriate way, whereas to produce new heavy particles, the energy is made as large as possible. All uses of electron colliders require very large luminosity (intensity).

Electromagnetic Force

See *Force* and *Electroweak Force.*

Electroweak Force
The descriptions of the electromagnetic and weak forces have been unified into a single description, the electroweak force. The electromagnetic and weak forces appear to be different because the W and Z bosons that mediate the weak force are massive, whereas the photon that mediates the electromagnetic force has no mass. Consequently, it is easier to emit photons than W and Z bosons. The electroweak unified theoretical description treats all the bosons on an equal footing and explains why they appear to be different, because of their masses.

Family
Quarks and leptons appear to come in three families, although only one family appears to be needed to explain the world we see. The other families differ only in that they are heavier. We do not yet understand why there are three families, but we can fully describe their behavior. This is one of the main mysteries of particle physics.

Fermion
Fermions are particles with half a unit of spin. They have different properties from particles with an integer unit of spin (bosons). Quarks and leptons, the matter particles, are fermions.

Feynman Diagrams
The rules of any quantum field theory can be formulated so that it is possible to draw a set of (Feynman) diagrams representing the processes that can occur, and to assign a probability of occurrence to the process represented by each diagram or set of diagrams. The diagrams are very helpful guides to thinking about what processes can occur. The structure of the theory determines which diagrams are allowed.

Field
Every particle is the origin of a number of fields, one for each nonzero charge the particle carries. Interactions occur when one particle feels the field of another (and reciprocally). There are electromagnetic fields, weak fields, and strong fields. Any particle with energy (remember that mass is a form of energy) sets up a gravitational field. In the Standard Model, particles get mass by interacting with a Higgs field, but the origin of the Higgs field is not yet understood.

Forbidden

Processes can be naïvely imagined that might occur, but should not occur according to the predictions of the Standard Model. Whether they occur is then a test of the Standard Model. If they occurred at the same rate as other processes the Standard Model would be wrong; if they occur at much smaller rates, or do not occur at all, they provide a clue as to how to extend the Standard Model. None of the processes forbidden by the Standard Model have been observed.

Force

All the phenomena we know of in nature can be described by five forces: the gravitational, weak, electrical, magnetic, and strong forces. The electrical and magnetic forces are unified into the electromagnetic force. Although the weak and electromagnetic forces appear different to us, they can be described as unified into one force (electroweak) in a more basic way; there is evidence that a similar unification of the electroweak force with the strong force also occurs. The attempt to unify all forces is an active research area. In particle physics, *force* and *interaction* mean essentially the same thing.

Gauge Boson

The strong, electromagnetic, and weak interactions are transmitted by the exchange of particles called gauge bosons (gluons, photons, and W's and Z's). The gauge bosons are the quanta of the strong, electromagnetic, and weak fields.

Gauge Theory

Any quantum field theory in which interactions occur between particles carrying charges, with strengths proportional to the sizes of the charges, and are transmitted by bosons that are quanta of the fields set up by the charges. In a gauge theory, once any particle exists (such as an electron) that carries any kind of charge (electric or weak or strong or any yet to be found), the associated boson that transmits the force must exist (photons or W and Z bosons or gluons or any yet to be found). Otherwise, the theory would not be consistent.

General Relativity

Einstein's theory of the gravitational interaction.

Gluino

The hypothetical supersymmetric partner of the gluon, differing only in that the gluino has spin 1/2, whereas the gluon has spin 1, and the gluino is heavier.

Gluon

The particle that transmits the strong (or color) force; the quantum of the strong field.

Gluon Jet

The color (or strong) force is so strong that colored particles (quarks and gluons) hit or produced in a collision can separate from other particles carrying color charge only by binding to one another and making color-neutral particles (hadrons), mainly pions. Thus an energetic gluon or quark becomes a narrow "jet" of hadrons as it moves along, turning its energy into the mass and motion of several hadrons. A quark or gluon appears in a detector as a jet of typically five to fifteen hadrons.

Grand Unification

The proposed unification of the weak, electromagnetic, and strong forces into a single force. This unification, if it occurs, must happen in the sense that the forces act as a single one at very short distances, a million billion times smaller than distances that have been studied experimentally so far; the forces behave differently when they are studied at larger distances.

Gravitational Force

See *Force*.

Graviton

The quantum of the gravitational field, which mediates the gravitational force.

Hadron

The properties of the color force and the rules of quantum theory allow certain combinations of quarks (and antiquarks) and gluons to bind together to make a composite particle; all such particles are called hadrons. When mainly three quarks bind, the resulting hadron is called a baryon. When quark and antiquark combine, the result is called a meson, and when gluons combine, it is called a glueball. Hadrons have diameters of about 10^{-13} cm. The proton and neutron are the most familiar baryons. Pions are the lightest mesons, so they are produced frequently in collisions. Kaons are the next lightest hadrons; their properties make them useful in many studies.

Helium Abundance

As the universe cooled after the Big Bang, it eventually reached a stage (about a minute after the beginning) when protons and neutrons formed, and then nuclei. Nuclei up to helium formed, but collisions were too soft for heavier nuclei to form. The theory of the Big Bang predicts the fraction of nuclei that are helium. That fraction has been measured, and the observed amount agrees very well with the predicted amount. This is one of the main reasons why it is generally believed that the universe began in a hot Big Bang.

Higgs Boson

The Higgs boson is the quantum of the Higgs field. If the theory implying the existence of Higgs bosons is correct, it will be possible to produce and detect Higgs bosons at the CERN LEP collider if they are light enough, and almost certainly at the Fermilab collider sometime between 2002 and 2006, depending on the mass of the Higgs boson and on how well the collider and detectors operate. The CERN LHC that is expected to begin operation in 2005 will be a Higgs boson factory. See also *Higgs Field, Higgs Mechanism, Higgs Physics,* and Chapter 7.

Higgs Field

In the Standard Model, particles (bosons and fermions) are thought to get mass by interacting with the Higgs field. The Higgs field and the way the particles interact with it must have very special properties for the masses to be included in the theory in a consistent way. The other fields we know of arise from particles that carry charges, but we do not yet understand the origin of the Higgs field. That is why physicists are very nervous about the Higgs physics and do not yet agree about whether Higgs bosons exist, even though there is good indirect evidence for the Higgs bosons. Many extensions of the Standard Model imply the existence of a Higgs field. See also *Higgs Mechanism.*

Higgs Mechanism

The Higgs mechanism is a special set of circumstances that must hold if bosons and fermions are to get masses from interacting with a Higgs field, given that the Higgs field exists. In the Standard Model these circumstances can be imposed, and in the supersymmetric Standard Model they can be derived. See Chapter 7 and Appendix A.

Higgs Physics

This is the combined physics that explains the origin of the Higgs field, the reason the Higgs mechanism applies, and the properties and study of the Higgs bosons.

Higgsino

The superpartner of the Higgs boson.

High-Energy Physics

Another name for particle physics, often used because much (though not all) of particle physics is based on experiments requiring high-energy beams.

Hot Dark Matter

See *Dark Matter.*

Inflationary Universe

According to the inflationary universe theory, as the universe expanded after the Big Bang, it went through a stage of very rapid expansion called inflation and then slowed down to the present rate of expansion.

Intensity

A measure of how often collisions occur at a collider. *See also* Luminosity.

Interaction

See *Force.*

Jet
See *Gluon Jet.*

Kaon
See *Hadron.*

Lagrangian
A Lagrangian is an equation that contains representations of all of the fundamental particles in the world and specifies how they interact. Given the Lagrangian, the rules of quantum theory specify how to calculate the behavior of the particles, how to build up all of the composite systems they form, and all the consequences of the basic theory.

Lepton
A class of particles defined by certain properties: leptons are fermions with spin 1/2 that do not carry color charge and that have another property called lepton number that is different for each family. The known leptons are the electron, the muon, the tau, and their associated neutrinos.

Lightest Superpartner
The superpartner with the least mass. The LSP may have several important roles. In particular, it may be the cold dark matter of the universe, and its properties are crucial for identifying the events of superpartner production at colliders, because all of the heavier superpartners decay into the lightest one.

Linear Electron Collider
Particles traveling in a curved path continuously radiate photons that carry away some of the particles' energy. The fraction of energy radiated increases with the energy of the particle, and the radiation happens with greater probability for lighter particles than for heavier ones. For electrons this loss of energy is a large effect at the circular CERN LEP collider, and it would be worse at a higher-energy collider, so it is unlikely that any future electron collider will be circular. The next electron collider built is expected to be a linear one, where the radiated energy loss is greatly decreased (NLC, for Next Linear Collider), modeled on the first linear collider, the SLC at SLAC.

Luminosity
Any collider has two basic figures of merit: the maximum energy it can supply to the collisions and how often it can cause collisions to occur. The number of events at a collider over some period of time is the product of two factors: the probability that something will happen if two particles actually collide, and the number of collisions. The latter is a property of the collider, not of the physics that governs the collision. It is called the luminosity. It depends on how many particles can be accelerated, how tightly bunches of them can be packed, and so forth. Loosely speaking we can refer to the luminosity as the intensity.

M-theory
The last, or nearly the last, stage on the way to the primary theory. See Chapter 9.

Mass
Mass is an intrinsic property of any object that measures how hard it is to make the object move. It can be thought of as weight, though the two are not quite the same (the mass of an object does not change, but if it were transported to a planet with a different mass, its weight would change).

Matter
It is useful to think of quarks and leptons as the basic particles that make up all the things around us, and of the photons and gluons that bind them as quanta of the fields. We call the quarks and leptons matter particles. Sometimes the term *matter particles* is used to mean fermions.

Matter Asymmetry
Our universe seems to be made of matter, but not antimatter, so there is an "asymmetry." Several ideas exist to explain how a universe could initially be symmetric, with equal numbers of protons and antiprotons, but then evolve into our asymmetric one, with about a billion protons for every antiproton, today. This is an active research area. Sometimes called the baryon asymmetry.

Maxwell's Equations
Electromagnetism, the unified theory of all electrical and magnetic phenomena, is summarized in a set of equations first written by Maxwell in

the 1860s. When they are extended to include the effects of the quantum theory, the theory of quantum electrodynamics (QED) is obtained. Physics students spend about a quarter of their time for two years learning how to solve Maxwell's equations, unless they plan to work in a sub-field that relies heavily on Maxwell's equations, in which case they spend much more time studying them.

Mediate
The effects of interactions are transmitted from one particle to another by exchange of particles called bosons. The bosons are said to mediate the interaction or force.

Meson
See *Hadron.*

Microwave Background Radiation
As the universe expanded and cooled, the original particles decayed or annihilated until only photons, neutrinos, protons, neutrons, and electrons that formed atoms remained. Today there is a cold gas of photons, about four hundred in each cubic centimeter of the universe, called the microwave background radiation, because the wavelength of the photons is in the microwave part of the spectrum. The properties of this background radiation can tell us a great deal about the properties of the universe and how it began, and they are the subject of intense study.

Missing Energy
When superpartners are produced at colliders, we expect each superpartner to decay into Standard Model particles plus the lightest superpartner, which interacts weakly and thus escapes the detector. The energy it carries off is expected to be one of the main signatures that tells us superpartners have been produced.

Molecules
Although atoms are electrically neutral, the positive and negative charges are not on top of one another, so there is some electric field outside an atom. Therefore, atoms can attract each other and form molecules, which can get very large.

Muon

A fundamental particle. A muon decays into an electron and neutrinos in about a millionth of a second. Muons are made in collisions at accelerators and in decays of other particles produced at accelerators and in cosmic ray collisions.

National Laboratories

Because much of the research in particle physics has to be done at large accelerators that are very expensive, the accelerators are built as national or international facilities at a few labs and are used by all particle physicists.

Neutrino

A fundamental particle. There is one neutrino for each of the three families of particles.

Neutron

Free neutrons have a lifetime of about fifteen minutes; they decay into a proton, an electron, and an antineutrino. When the neutrons are bound into nuclei (such as those in us), the decays are no longer possible because of subtle effects explained by quantum theory, so the neutrons in nuclei are as stable as protons. See also *Hadron*.

Newton's Constant G

Newton's law of gravitation says that the gravitational force between any two bodies is proportional to the product of their masses and decreases as the square of the distance between them. This statement is turned into an equation by inserting the constant G, so the force $F = Gmm'/r^2$. Because all of the particles feel the gravitational force, G is universal, so G can be used to form quantities with dimensions, giving the Planck scale. G is measured by finding the force between two objects of known masses separated by a known distance.

Newton's Laws

Newton formulated the law that describes the gravitational force (see *Newton's Constant G*), and three laws that describe motion. The first law says that every moving body moves in a straight line at constant speed unless a force acts on it. The second law says that the product of the mass of a body and its acceleration is equal to the force acting on it ($F = ma$). The third law says that when one body applies a force on a second, the second applies an equal and oppositely directed force on the first.

Nuclear Force

Although protons and neutrons carry no strong (or color) charge, at tiny distances near a proton or neutron, the cancellation of the strong field from its constituent quarks and gluons is incomplete, leaving a residual strong force that leaks outside the proton and neutron. This is the nuclear force that binds protons and neutrons into nuclei.

Nucleus

Although protons and neutrons are color-neutral composites of quarks and gluons, the quarks and gluons are not all at the same places, so some of their color (or strong) fields exist outside the proton or neutron, giving an attractive force that binds protons and neutrons into nuclei. The attractive effects of this residual color force are offset by the electrical repulsion of the protons, so nuclei with too many protons cannot exist. It turns out that there are ninety-two stable or long-lived nuclei in nature. They are the nuclei of the atoms of the ninety-two chemical elements.

Particle

The term *particle* is used somewhat loosely and includes not only the elementary quarks and leptons and bosons, but also the composite hadrons. It also includes any (currently hypothetical) new particles that might be discovered, such as the supersymmetric partners of the quarks and leptons and bosons.

Particle Physics

Those engaged in this field of physics study the particles and try to understand their behavior and properties. Sometimes a distinction is made between the study of quarks, leptons, gauge bosons, and Higgs physics, and the study of hadron physics, which aims to relate the properties of the hadrons to the theory of the color force. More broadly, the goals of particle physics are to understand not only the description of the particles and their interactions but also why the laws of nature are what they are and how the universe arises from those laws.

Photino
The supersymmetric partner of the photon.

Photon
The photon is the particle that makes up light. It transmits the electromagnetic force. It is the gauge boson of electromagnetism. Once electrons exist, quantum theory implies that photons must exist and must have the properties they do.

Pion
The lightest hadron, and therefore the one most often produced in collisions. See also *Hadron*.

Planck Energy, Length, Time
See *Planck Scale*.

Planck's Constant h
Many things are quantized, such as the energy levels of atoms. Planck's constant, h, sets the scale of quantization: Energy levels are separated by amounts proportional to h, the amount of spin a particle can have is a multiple of h, etc. Planck originally found that the energy radiated by a heated body was emitted in quanta, rather than emission of any continuous amount being possible. The amount of energy was always an integer multiple of hf, where f is the frequency or color of the radiation; that is, hf of energy can be emitted, or $2hf$, or $3hf$ and so on, but not amounts in between. By separately measuring the frequency, Planck deduced the value of h, which is 6.63×10^{-34} joule-second. See also *Planck Scale*.

Planck Scale
The Planck scale refers to certain values of length, time, and energy or mass. To understand how these values originate, suppose you were trying to explain to an intelligent being in another galaxy how long humans typically lived. You couldn't use hours or years, because those units are defined on earth (for example, by how long it happens to take our planet to go around its sun once), so a being in another galaxy wouldn't know what you meant. However, every physicist in the universe knows the values of Planck's constant (h), the speed of light in vacuum (c), and the universal strength of the gravitational force (G). You could use those values to form

ratios that define a universal unit of time called the Planck time and then tell the being from another galaxy our typical lifetime in units of Planck times. Similar units for length and mass or energy can be defined. Max Planck realized this possibility and defined these units at the beginning of the twentieth century. Because the Planck scale units are the only universal ones, we expect the fundamental laws of nature to be simple in form when expressed in those units. See Chapter 3.

Point-Like

If matter is probed with projectiles that are large and have energies that are less than what is needed to change the energy levels of an atom, then atoms will seem to be point-like objects. If the energy is increased, eventually the projectile will penetrate the atom but will encounter the nucleus, which will seem to be point-like. With higher energy, the nucleus will appear to be made of point-like protons and neutrons. With still higher energies, the protons and neutrons will be seen to be made of point-like quarks and gluons. As the energies of projectiles were increased still more, quarks and leptons might have been seen to be made of something still smaller, but that has not happened. Rather, they behave as point-like up to the highest energies they have been probed with—energies well beyond those for which we would have expected to find more constituents if history were to repeat itself once more. Further, the structure of the Standard Model theory suggests that quarks and leptons are the fundamental, point-like constituents of matter.

Positron

The antiparticle of the electron.

Predict

Predict is used in the normal sense that a theory may predict some unanticipated or as-yet-unmeasured result. It is also used in another sense: A theory can be said to predict a result that is already known, because once the theory is written, it gives a unique statement about that result. Sometimes an in-between situation holds, in that the theory predicts a result uniquely in principle, but the prediction depends on our knowing some other quantity or requires very difficult calculations.

Primary Theory

The name used in this book for the theory sought by many particle physicists that includes not only the Standard Theory but also the theory of gravity, explains why the primary theory itself takes the form it does, explains what quarks and other particles are, explains what space and time are, and more. See *Theory of Everything.*

Projectile
To study particles and their interactions, it is necessary to probe them with projectiles. The projectiles are other particles (electrons, photons, neutrinos, and protons) because these are small enough and can be given enough energy.

Proton
See *Hadron.*

Proton Decay
If the Standard Model were the complete theory that described nature, protons would be stable, never decaying. If the Standard Model is part of a more comprehensive theory that unifies quarks and leptons, then possibly protons are unstable, though with extremely long lifetimes. Some basic theories imply that protons decay; others do not. Experiments that search for proton decay are very important, because knowing that it occurred (and what the proton decayed into) would provide valuable information about how to extend the Standard Model.

Quanta
Each particle is surrounded by a field for each of the kinds of charges it carries, such as an electromagnetic field if it has electric charge. In the quantum theory, the field is described as made up of particles that are the quanta of the field. More loosely, the smallest amount of something that can exist.

Quantum Field Theory
When interactions among particles are described as transmitted via the exchange of bosons, the methods of quantum field theory are used.

Quantum Theory
The quantum theory provides the rules with which to calculate how matter behaves. Once scientists specify what system they want to describe and what the interactions among the particles of the system are, then the equations of the quantum theory are solved to learn the properties of the system.

Quark

A fundamental particle. Quarks are very much like electrons, but they also carry strong charge and thus have another interaction, one that can bind them into protons and neutrons. There are six quarks, called up (u), down (d), charmed (c), strange (s), top (t), and bottom (b).

Quark Jet

Because quarks must end up in hadrons, quarks that are produced in collisions actually appear in detectors as a narrow jet of hadrons, mostly pions. See also *Gluon Jet*.

Radioactive Decay

Some nuclei are unstable but live long enough to exist as matter until they decay. When they decay, they can emit several particles: photons, electrons, positrons, neutrinos, neutrons, and even helium nuclei. For historical reasons, such decays are called radioactive decays. Sometimes scientists use the emitted particles as tools to do experiments.

Reductionist

One way to study the natural universe is to study very detailed aspects of nature, to take things apart and see what they are made of, and to focus on small steps. This approach is called reductionist. It has been a powerful success, enabling us to build up the remarkably complete description of nature we now have. Whenever possible, scientists have tried to unify subfields as they became understood. Recently in particle physics, the trend toward unification has been increasingly successful. For physicists, reductionism includes the associated unification. See also *Unification*.

Relativistic, Relativistic Invariance

Whenever particles can move at speeds near the speed of light, and whenever fields are involved, the description of nature must satisfy the requirements of Einstein's "special relativity" theory.

RIP

An acronym I use to emphasize that some of the subjects we cover are "Research in Progress," whereas others are well-established areas that may be extended but will not be significantly modified (such as the Standard Model).

Rules

In order to have a complete understanding of nature, it is necessary to know the particles, the forces that determine the interactions of the particles, and the rules for calculating how the particles behave. For the motion of objects normally on earth or in the sky, the rule to use to calculate the behavior of particles is Newton's second law, $F = ma$. When atomic or smaller distances are involved, the Schrödinger equation of quantum theory replaces Newton's second law. In particle physics, additional relativistic requirements are added to make the complete set of rules: quantum theory and Einstein's special relativity. See Chapter 1.

Schrödinger Equation

The equation from quantum theory that tells how to calculate the effects of the forces on the particles. It is the quantum theory equivalent of Newton's second law.

Science

Science can be defined as a self-correcting way to get knowledge about the natural universe, plus the body of knowledge obtained that way. It is both a method and the resulting understanding and knowledge. The method requires making models to explain phenomena, testing them experimentally, and revising them until they work. The goal of science is understanding. Once part of the natural world is understood, it may be possible to develop applications of the new knowledge. The process of developing such applications is properly called technology, not science. Although scientific knowledge may, and usually does, lead to technology, science is not necessary for technology, and technological developments have led to new science as much as the opposite. Before the time of Galileo, many technological developments occurred that had no scientific connection. Since the time of Maxwell and his formulation of the electromagnetic theory, nearly all technological developments have depended on earlier science. In recent years the words *science* and *technology* have been frequently misused, as though they were interchangeable. Because science and technology are really different, it is better to distinguish carefully between them.

Selectron

The supersymmetric partner of the electron.

Signature

A new particle will have some characteristic behavior in a detector that allows it to be recognized. Particles that decay into others do so in a unique way that is different for every kind of particle. Knowing the properties of the particle allows us to calculate how it will decay. The features that allow a new particle to be identified in a detector are called its signature.

Slepton

The supersymmetric partner of any of the leptons.

Smatter

The superpartners of the Standard Model particles. This book argues that the experimental discovery of smatter will provide us with information that will be essential for gaining insights into the ultimate laws of nature, the primary theory.

Solar Neutrinos

The reactions that fuel the sun lead to the emission of photons, which reach the earth as sunlight, and of neutrinos, which we do not see with our eyes but which can be detected in special neutrino detectors. At present there is great interest in these neutrinos, because the number being detected is fewer than expected, and this may be a signal that neutrinos have mass, in which case we could account for the lesser number detected. If they have mass, the experiments to detect them will allow the value of their mass to be measured.

Special Relativity

The constraints of special relativity are two conditions that Einstein pointed out should be satisfied by any acceptable physical theory. Somewhat oversimplified, these conditions are, first, that light moves at the same speed in vacuum regardless of how it is emitted and, second, that scientists working in different labs moving with different relative speeds should formulate the same natural laws. The constraints imposed by these conditions have surprising implications for the structure of acceptable theories. For example, the Schrödinger equation of quantum theory does not satisfy these conditions. But when it was generalized by Dirac to do so, the resulting equation led to the prediction of antiparticles, which need not have existed from the point of view of quantum theory alone.

Spectra

Atoms can exist in a number of discreet energy levels. They emit or absorb photons when they make transitions from one level to another. The energies of the photons emitted or absorbed by one atom are different from those of all other atoms. The photon energies are directly related to their frequencies, which set their colors in the spectrum, so by observing the colors of the photons, it is possible to determine which atoms are being observed. This can be done in a laboratory, and it can also be done with the light reaching us from stars, near or distant, which enables us to identify the atoms that stars are made of. Only the same ninety-two elements we find on earth are seen throughout the universe.

Speed of Light

Light and all other massless particles travel in vacuum with a speed, usually labeled c, whose value is about three hundred million meters a second. Special relativity implies that no particle or signal can move faster than the speed of light and that photons always have this speed, regardless of the speed of their source.

Spin

Spin is a property that all particles have. It is as though particles were always spinning at a fixed rate (which could be zero), which can be different according to the type of particle. It is not quite right to think of them actually spinning, because the particles do not have to have spatial extension to have spin; calling this property spin is an analogy. The amount of spin is required by the quantum theory to come in definite amounts; if the unit is chosen to be Planck's constant, h, divided by 2, then particles can have zero spin, half a unit of spin, one unit of spin, etc.

Spontaneous Symmetry Breaking

Often the equations of a theory may have certain symmetries, though their solutions may not; the symmetries are hidden, or broken. For example, the equations may describe several particles in identical ways, so the equations are unchanged if the particles are interchanged, but the solutions may give the particles different properties. (A simple example is given in Chapter 1.) This phenomenon is called spontaneous symmetry breaking.

Squark

The supersymmetric partner of any of the quarks.

Stable Particle
Particles that do not decay into others. See also *Decay*.

Standard Model
The very successful theory of quarks and leptons and their interactions that is described in this book is called the Standard Model by particle physicists. The name arose historically as the theory developed and then was difficult to change because it is widely used. The Standard Model is the most complete mathematical theory of the natural world ever developed and is well tested experimentally.

String Theory
String theory is a theory that aims to unify all of the forces and particles of nature and explain why they are as they are. In string theory, there is only one force (gravity), in ten space-time dimensions, but when looked at from our four-dimensional world, the extra dimensions imply the other forces we observe. Particles are strings that vibrate in different ways to account for their various properties. String theories appear to allow the construction of a quantum theory of gravity. String theory is RIP.

Strong Force
See *Force*.

Structure
Objects have structure if they have parts—that is, if they are made of other things. Whether objects have structure can be learned from experiments that probe them with projectiles. Over the past century, each stage of matter that was found as it became possible to search for ever-smaller things turned out to have structure. Quarks and leptons appear not to have structure, so perhaps the search for the basic constituents has finally ended. There are also theoretical arguments that quarks and leptons are the basic constituents.

Subatomic Particle
Any particle that is contained in an atom, or any particle that can be created in collisions of such particles, is loosely called subatomic, whether it is composite like a proton or elementary like a quark or electron.

Superpartner
If the theory that describes nature has a symmetry called supersymmetry, then every normal particle (the ones we know) has associated with it a partner that differs only in its spin and its mass.

Superspace
Supersymmetry can be formulated in several ways. One is to imagine associating another coordinate that has special properties with each of our normal spacetime coordinates, giving a kind of space called superspace. Writing theories in superspace makes them supersymmetric. This way of constructing supersymmetric theories is harder to picture than associating superpartners with each Standard Model particle, but it leads to the same results and sometimes facilitates deriving mathematical properties of the theories.

Superstring
String theories are expected to be supersymmetric and are often called superstring theories.

Supersymmetry
A hypothetical symmetry that describes nature and says that even though fermions and bosons seem to us to be very different in their properties and their roles, in the theory itself they appear in a symmetric way. If supersymmetry is indeed realized in nature, then every particle has a superpartner.

SUSY
A common abbreviation for supersymmetry.

Technology
See *Science.*

Theory
The word *theory* is usually used precisely in physics. Theories are not conjectures but sets of equations whose solutions describe physical systems and their behavior.

Theory of Everything
A "theory of everything" would not only describe how things work but also explain why things are the way they are. The name is unfortunate in

one way, because it does not tell how to deduce the behavior of complex systems from a knowledge of their components. In this book I have used the name *primary theory* instead.

Transmit
See *Mediate*.

Uncertainty Principle
The uncertainty principle is a consequence of quantum theory. It implies that a pair of observables cannot both be measured simultaneously to arbitrary accuracy. It can often be used to understand quantum theory results in a simple way.

Unification
Scientists have sought for centuries to unify the descriptions of apparently different phenomena by showing that they were due to the same underlying natural laws and that complex levels of matter were made of simpler levels. This unification process is a subject of very active research about the forces of nature today. The possible unification of the strong, electromagnetic, and weak forces is called a grand unification. There is a continuing effort to unify these forces with gravity. String theories seem to do that successfully.

Unstable Particle
See *Decay*.

Vacuum
Any physical system will settle into the lowest-energy state it can, which in particle physics we call its vacuum state. For most systems, this is the state where the fields making up the system are zero, but theorists hypothesize that for systems containing Higgs fields, the lowest energy occurs when the Higgs field takes on a constant value different from zero. The value of the Higgs field in that system is called its vacuum expectation value.

Vacuum Expectation Value
See *Vacuum*.

Weak Charge
See *Charge*.

Weak Force
See *Force.* The weak force is described in Chapter 4.

Wino
The supersymmetric partner of the W boson.

Zino
The supersymmetric partner of the Z boson.

Index